Climate Change and Sustainable Development

Climate Change and Sustainable Development
Mitigation and Adaptation

Dalia Štreimikienė and Asta Mikalauskienė

CRC Press
Taylor & Francis Group
Boca Raton London New York

CRC Press is an imprint of the
Taylor & Francis Group, an **informa** business

First edition published 2021
by CRC Press
6000 Broken Sound Parkway NW, Suite 300, Boca Raton, FL 33487-2742

and by CRC Press
2 Park Square, Milton Park, Abingdon, Oxon, OX14 4RN

CRC Press is an imprint of Taylor & Francis Group, LLC

ISBN: 978-0-367-55031-8 (hbk)
ISBN: 978-0-367-55033-2 (pbk)
ISBN: 978-1-003-09164-6 (ebk)

Typeset in Palatino
by codeMantra

Contents

Preface

Today's climate change causes the biggest risks to world population. The climate change and sustainable development issues are closely interlinked, and policies targeting these interlined issues need to be harmonised as well. Climate change also provides a lot of opportunities for business, thus addressing climate change mitigation and adaptation policies. This book deals with these important issues and provides valuable input in the field of climate change research area. The aim of this book is to present climate change threats to sustainable development, to provide policies of climate change mitigation and adaptation, and to assess their impact on international trade and business development perspectives.

The book deals with the challenges of climate change and discourses the role of climate change mitigation and adaptation policies in addressing these challenges through implementing sustainable development goals. The linkages between climate change and sustainable development are very important as climate change is the main environmental challenge to sustainable development and climate change mitigation and adaptation policies have serious implications on other economic and social issues of sustainable development such as hunger, good health and well-being, economic growth, industry and innovation, sustainable cities and communities, clean and affordable energy, and responsible consumption and production. Therefore, the book discusses the centrality of climate change in terms of achieving all Sustainable Development Goals set by Agenda 2030. Thus, climate change mitigation and adaptation policies should be developed and assessed by taking into account sustainable development issues as well.

The structure of this book provides logical background for understanding the aforementioned linkages between climate change and sustainable development. The book introduces causes and consequences of climate change and provides international climate change regulation efforts including recent Paris Agreement and country commitments. The following part of the book presents the recent trends of globalisation and its impact on climate change challenges and opportunities for business and world countries in terms of international trade. The subsequent parts of the book introduce the most important climate change mitigation and adaptation tasks and policies for world countries. The theoretical background of economic measures to address climate change is provided with the focus of greenhouse gas (GHG) emission trading schemes. The case of EU Emission Trading Scheme and its implications are throughout discussed. Climate change adaptation issues are discussed by providing the most important threats of climate change for developing world countries and actions necessary to adapt to climate change in various parts of the world.

The book is based on research studies on various climate change issues conducted by book authors during recent decade covering energy, transport, agriculture, etc. sectors.

The book is constituted of five interconnected parts that sequentially reveal the principal provisions of climate change, sustainable development, globalisation, and mitigation and adaptation policies and their correlations.

In Part 1 of the book, the climate change phenomena and its causes and consequences are provided including GHG emission sources and trends. The main legal climate change regulation issues set by international institutions are addressed in this part as well.

In Part 2 of the book, the concept of sustainable development is introduced by highlighting the sustainable development goals and linkages between climate change and sustainable development challenges.

In Part 3 of the book, globalisation and climate change issues are addressed. International trade liberalisation impacts on climate change are analysed, and the business development opportunities are provided based on analysis performed. Good practice case studies for business development in selected world countries are also presented.

In Part 4 of the book, the in-depth analysis of climate change mitigation is provided. The main attention is being paid on economic and flexible climate change mitigation tools. The practical example of EU GHG emission trading scheme and its implications are discussed.

In Part 5 of the book, climate change mitigation issues are discussed by highlighting importance of climate change adaptation for developing countries and migration trends development of cities and agriculture.

Author

Prof. Dr. Dalia Štreimikienė is a professor at Vilnius University, Kaunas faculty and Leading Research Associate at Lithuanian Energy Institute, Lithuania. Her main areas of research include sustainable development, sustainability assessment, and corporate social responsibility. She is an author of more than 100 papers in international journals referred at WoS. H index 19. She is an invited Editor for several special issues of Sustainability journal and Editor-in-Chief of international journal *Transformations in Business and Economics* referred at WoS. Orcid: 0000-0002-3247-9912. https://www.researchgate.net/profile/Dalia_Streimikiene2.

Prof. Asta Mikalauskienė is a professor at Vilnius University, Kaunas faculty, Lithuania. Her main areas of research include sustainable development, sustainability assessment, and corporate social responsibility. She is an author of many papers in international journals referred at WoS. Orcid: https://orcid.org/0000-0002-4301-2058. https://www.researchgate.net/profile/Asta_Mikalauskiene.

Introduction

Gradual changes in nature, such as the greenhouse effect caused by exhaust gas emissions and the increase in average temperature followed by sea-level rise, are currently gaining momentum and steadily intensifying. The planet is undergoing irreversible processes when the environment around us is no longer able to absorb pollution, generated by human activity, which is constantly emitted into the atmosphere. The global economy, based on market laws, and the liberalised international trade increase the operational risk in terms of the survival of natural environment and life. For many years, humanity has denied climate change as a false rumour by appealing to the intellectual capacities of individuals that will allow creating technical miracles having the ability to successfully absorb carbon dioxide from the atmosphere. The long-prevailing and still dominant view that the best protection against the consequences of possible climate change is the maximum focus on economic development and generated profits shows our attempt to concentrate on rationality. However, the created systems push us towards an existential crisis of humanity, as a species of life, which requires little desirable social, political, and economic change to overcome it. For a long time, our human nature has been holding us back from carrying out activities that could contribute to reducing climate change because threats have always seemed too distant.

This book deals with the challenges of climate change and discourses the role of climate change mitigation and adaptation policies in addressing these challenges through implementing sustainable development goals. The linkages between climate change and sustainable development are very important as climate change is the main environmental challenge to sustainable development and climate change mitigation and adaptation policies have serious implications on other economic and social issues of sustainable development such as hunger, good health and well-being, economic growth, industry and innovation, sustainable cities and communities, clean and affordable energy, and responsible consumption and production. Therefore, the book discusses the centrality of climate change in terms of achieving all Sustainable Development Goals set by Agenda 2030. Therefore, climate change mitigation and adaptation policies should be developed and assessed by taking into account sustainable development issues as well. The important input of this book is presentation of clear linkages between climate change mitigation and adaptation policies and implementation of sustainable development goals by presenting framework for harmonisation of climate and sustainable development policies and discussing their mutual outcomes.

The structure of the book provides logical background for understanding the aforementioned linkages between climate change and sustainable development. The book introduces causes and consequences of climate change and provides international climate change regulation efforts including recent Paris Agreement and country commitments. The following part of the book presents the recent trends of globalisation and its impact on climate change challenges and opportunities for business and world countries in terms of international trade. The subsequent parts of the book introduce the most important climate change mitigation and adaptation tasks and policies for world countries. The theoretical background of economic measures to address climate change is provided with the focus of GHG emission trading schemes. The case of EU Emission Trading Scheme and its implications are throughout discussed. Climate change adaptation issue are discussed by providing the most important threats of climate change for developing world countries and actions necessary to adapt to climate change in various parts of the world.

Therefore, the aim of this book is to present the main climate change threats to sustainable development and to analyse policies of climate change mitigation and adaptation from economic point of view by addressing globalisation, international trade, and business development opportunities and challenges. The book is based on research studies on various climate change issues conducted by book authors during recent decade covering energy, transport, agriculture, etc. sectors.

The book is constituted of five interconnected parts that sequentially reveal the principal provisions of climate change, sustainable development, globalisation, and mitigation and adaptation policies and their correlations.

In Part 1 of the book, the climate change phenomena and its causes and consequences are provided including GHG emission sources and trends. The main legal climate change regulation issues set by international institutions are addressed in this part as well.

In Part 2 of the book, the concept of sustainable development is introduced by highlighting the sustainable development goals and linkages between climate change and sustainable development challenges.

In Part 3 of the book, globalisation and climate change issues are addressed. International trade liberalisation impacts on climate change are analysed, and the business development opportunities are provided based on analysis performed. Good practice case studies for business development in selected world countries are also presented.

In Part 4 of the book, the in-depth analysis of climate change mitigation is provided. The main attention is being paid on economic and flexible climate change mitigation tools. The practical example of EU GHG emission trading scheme and its implications are discussed.

In Part 5 of the book, climate change mitigation issues are discussed by highlighting importance of climate change adaptation for developing countries and migration trends development of cities and agriculture.

This book is useful to academic community and policymakers as well as business community, as it presents theoretical frameworks for climate change mitigation and adaptation policies development and provides useful policy examples in specific world countries and also reveals business development opportunities linked to these policies based on selected case studies.

1

Causes of Climate Change and Legal Regulations

1.1 Causes of Climate Change

Since the end of the 19th century, the greatest impact on the climate has been achieved by the increasing atmospheric pollution caused by human economic activity: the concentration of greenhouse gases (GHGs) and ozone-depleting gases is rising. These climate change factors are transforming the state of the entire climate system at an unusually fast rate.

Aerosols of volcanic and anthropogenic origin cause the opposite effect – cooling – since they reduce the transparency of the atmosphere and the solar radiation inflow to the Earth's surface.

Other global processes and reversible effects of the climate system are also significant to the climate sphere. These processes and effects include the rapid desertification of periodically wet savannas and scrubs due to overgrazing, wind erosion, and disturbed periodicity of seasonal rains as well as the decrease of stratospheric ozone.

The Earth's atmosphere has a natural physical property to warm up and retain the heat. Because of this phenomenon, most of the solar energy, received by the Earth during the day, is used to warm the atmosphere rather than the surface. At night, when the Sun is on the other side of the horizon, the warm air of the atmosphere cools slowly and steadily. This phenomenon of heat accumulation and retention is one of the necessary conditions for sustaining life on Earth. It is usually called the greenhouse effect (IPCC 2007). Planets with no atmosphere do not have this so-called greenhouse effect. During the day, solar energy burns the surface of these planets and, when the Sun turns beyond the horizon, the temperature immediately becomes extremely low.

The day and night temperatures of Mercury, which does not have an atmosphere, are 400°C and –160°C, respectively. The temperature of Venus, the dense atmosphere of which consists of 96% CO_2, does not fluctuate and is 462°C. Even though Venus orbits the Sun at the distance two times greater than Mercury, its average temperature is higher than the Mercury's maximum. Thus, the temperature of the planets is determined not only by the distance from the Sun but also by the ability of their atmospheres to absorb the heat it emits.

Earth's atmosphere is composed of a lot of different gases, and not all of them have the same property of warming up. For some fundamental reasons, symmetrical molecules (N_2, O_2, Ar, etc.) do not hold heat and quickly radiate it into space. Meanwhile, asymmetrical molecules (H_2O, CO, CO_2, CH_4, SO_2, O_3, etc.) retain heat but to different extents. A great part of the heat is retained in the atmosphere by water vapour molecules. Without this vapour, the average temperature of the whole Earth would be –16°C instead of the current 14°C. It would also be falling significantly faster and lower at night.

GHGs are gaseous components of the atmosphere of natural and anthropogenic origin, resulting in the so-called "greenhouse effect". Atmospheric GHGs absorb a part of long-wave infrared (IR) radiation of Earth's surface and return it towards the surface in a form of atmospheric counter-radiation; due to this, the temperature of lower tropospheric layers rises. A similar effect forms in an ordinary greenhouse of gardeners (therefore, it is called "the greenhouse effect"); only in this case, the role of GHGs is performed by glass or polyethylene film. Short-wave solar radiation penetrates the greenhouse; however, IR radiation emitted by the warmed soil is retained – the temperature in the greenhouse rises.

The main GHGs are:

- Water vapour
- Carbon dioxide (CO_2)
- Methane (CH_4)
- Nitrous oxide (N_2O)
- Sulphur hexafluoride (SF_6)
- Hydrofluorocarbons (HFCs)
- Perfluorocarbons (PFCs)
- Chlorofluorocarbons (CFCs)

If the atmosphere did not create a greenhouse effect, the average annual temperature of our planet would be –18°C, while now it is +15°C. It is obvious that as the concentration of GHGs increases, the greenhouse effect intensifies and the temperature of lower tropospheric layers rises.

The largest contributor in the creation of greenhouse effect is water vapour (21°C), then follows carbon dioxide (7°C), tropospheric and stratospheric ozone (2°C), nitrous oxide (1.4°C), and methane (0.8°C); all remaining GHGs also coincide with 0.8°C. Only the amount of water vapour is not directly affected by human economic activity; nevertheless, it should be kept in mind that as the troposphere warms, the amount of water vapour is also increasing: the so-called positive feedback effect occurs.

The concentration of GHGs in the atmosphere can be expressed in parts per million (ppm) or in parts per billion (ppb). 1 ppm means one molecule of a given gas per 1,000,000 of all existing gas molecules.

Since the Industrial Revolution in the 19th century, the concentration of GHGs in the atmosphere has been steadily rising. The concentration of CO_2 has mainly increased because of the fossil fuels (coal, gas, oil) combustion in the energy and transport sectors, various industrial technological processes, and deforestation (i.e., removal of the GHG sinks). Methane is emitted by extracting, transporting, and using natural gas and coal, as well as by animal husbandry, rice fields, and landfills. Nitrous oxide is formed by using nitrogen fertilisers and producing synthetic fibre.

HFCs, PFCs, and SF_6 are used as alternatives to ozone-depleting CFCs which, according to the Montreal Protocol, should be phased out.

However, GHGs do not absorb IR radiation of the Earth's surface in an equally efficient manner. Some of them absorb different IR wavelengths, while others absorb only narrow-spectrum waves. Moreover, GHGs greatly vary in terms of the duration of their persistence in the atmosphere. For example, CO_2 can last from 5 to 200 years in the atmosphere, methane can last for 12 years, nitrous oxide can last for 114, while some of the fluorinated GHGs (SF_6, CF_4) can last even for tens of thousands of years. Therefore, one tonne of GHGs emitted into the atmosphere has a different impact on the greenhouse effect, i.e., it has a different potential for creating a greenhouse effect. If the above-mentioned CO_2 potential equates to 1, within 100 years (regarding the duration of its persistence in the atmosphere), the global methane potential will be 23 times greater than that of CO_2, 296 times than nitrous oxide, 22,200 times than SF_6, from 12 to 12,000 times than various HFCs, and 5,700–11,900 times than PFC. Even if global GHG emissions were suddenly reduced drastically, it would take several centuries until the concentration of these gases in the air would reach the pre-industrial level.

In its Third Assessment Report, the Intergovernmental Panel on Climate Change (IPCC) provided new and obvious evidence that current climate warming has started mainly as a result of anthropogenic activities. Since the middle of the 18th century, the concentration of CO_2 in atmosphere increased by 31% (from 280 to 375 ppm), methane by 150% (from 700 to 1,750 ppb), and nitrous oxide by 18% (from 270 to 320 ppb), while the various HFCs and PFCs did not exist in the atmosphere until the middle of the 20th century (their emissions began with the development of the chemical and electronic industries). The current concentration of CO_2 is the highest in the last 420,000 years.

GHG emission from the process of human economic activities is increasing, while the areas of tropical rainforests and temperate coniferous and mixed forests are declining rapidly; therefore, less and less carbon dioxide is included in the photosynthesis. These trends are most occurred in developing countries in Latin America, South and South-East Asia, and Central Africa, but are also relevant for developed countries.

The IPCC reports admit that unless strict measures are taken to reduce these emissions, in comparison with the pre-industrial period, GHG concentration will double by the middle of this century.

To prevent drastic changes in social sphere and ecosystems, it is necessary to ensure that global temperature shall not rise more than 2°, while the GHG concentration, expressed in CO_2 equivalent, shall not exceed 550 ppm (now it is 425 ppm).

GHG accounting (inventory) is carried out in accordance with the IPCC methodology. GHG emissions are calculated using established emission coefficients and statistics of individual technologies or branches of the economy.

GHG emissions from energy, industrial processes, usage of solvents and other volatile substances, agriculture, changes of land use, and waste management are calculated. The inventory groups and subgroups of GHG sources are provided in Table 1.1.

The energy and transport sectors are dominated by carbon dioxide. These sectors cover the entire cycle of fuel use: extraction, transportation, storage, distribution, recycling, and combustion. Combusted fuel and GHG emissions from aviation, rail, maritime, and motor transport used in international transportation are included in the balance of the country that is the owner of these vehicles.

The agricultural sector is dominated by methane gas resulting from fermentation processes in the gastrointestinal tract of livestock as well as from animal husbandry waste, manure, and biogas management. Nitrogen fertilisers, used in agriculture, cause the release of nitrous oxide from the soil.

Changes in land use can have a twofold impact on the amount of GHGs. As the area of forests decreases, carbon dioxide consumption for photosynthesis also diminishes, which causes the increase of GHGs in the atmosphere. On the other hand, newly grown forests as well as biennial and perennial agricultural crops stimulate the removal of carbon dioxide from the atmosphere.

Waste management is a typical source of methane formation. Anaerobic bacteria break down organic waste materials into methane. Due to similar processes, methane is also released during wastewater treatment processes.

1.2 The Consequences of Climate Change

Climate change is happening here and now. Scientists are certain that the consequences of climate change are evident on every continent and in every ocean. Both the human and nature suffer from them. The number and frequency of outbursts of diseases are increasing in Africa, the number of fires

TABLE 1.1

Accounting Groups and Subgroups of Greenhouse Gas Sources

Total Emissions and Sinks
1. **Total energy**
1.A. Fuel combustion
1.A.1. Energy and its transformation
1.A.2. Industry
1.A.3. Transport
1.A.4. Central heating
1.A.5. Local heating
1.A.6. Agriculture
1.A.7. Other areas
1.B. Volatile fuel emissions
2. **Industrial processes**
2.A. Inorganic chemicals
2.B. Organic chemicals
2.C. Manufacture of cement and lime
2.D. Other areas
3. **Solvents and other used materials**
3.A. Paint
3.B. Degreasing and dry cleaning
3.C. Manufacture/use of chemicals
3.D. Other areas
4. **Agriculture**
4.A. Release of stomach gas
4.B. Livestock manure
4.C. Agricultural land
5. **Change of the purpose of the land and forestry**
5.A. Alternation in other forest biomass resources
5.B. Changes of the forest and grassland/pasture
5.C. Abandoned lands
6. **Waste**
6.A. Landfills
6.B. Wastewater

that cause losses and take many lives is rising in North America, while agricultural productivity and food quality are declining in South America.

The outcome of recent ravaging floods and other extreme weather phenomena that have damaged the planet indicate that we are still not ready for climate change. The report (IPCC 2018) claims that "adaptation deficit" is felt not only in developing countries but also in developed regions.

The majority of the world's ecosystems are already damaged. Many species are doomed to become extinct because they will not manage to adapt to quickly changing conditions. The degradation of coral reefs, caused by ocean acidification, warming, and eutrophication, has become a serious problem. Fish that depend on corals are quickly disappearing, their diversity, and abundance are decreasing. Even warming of 2°C is a very serious threat to fish resources.

Global temperature has already risen by 0.85°C, and only urgent and ambitious actions to reduce atmospheric pollution can save us from an increase greater than 2°C. Human GHG emissions are not decreasing, and until that happens, the risk of climate change consequences will only increase.

Europe is forecasted to face serious economic and social losses due to heatwaves, which will lead to increased disease rates, reduced harvests, and forest fires. Lesser resources of groundwater and river water together with the growing demand will increase the risks of water scarcity. Only the risk of economic and social losses, caused by floods, is of lesser extent.

The greatest threat in North America will be caused by fires. In Africa and Asia, this is caused by droughts and heatwaves. Meanwhile, the inhabitants of Central and South America will suffer from floods, whereas inhabitants of polar regions will suffer from the consequences of glacial melting. Increasingly serious damages will be inflicted to the fisheries sector, which belongs to the extremely high-risk zone. Extreme meteorological phenomena will also increase in frequency, severity, and duration. If the average global temperature rises by more than 4°C, casual activities, such as growing food or working outdoors, in many areas where people currently live, will become impossible due to the excess heat. Climate change will determine higher food prices and agricultural market volatility, which, in addition, can reduce food access. The provided data shows that climate change will cause the agricultural output to decline by 2%–3% each decade. There is a high probability that harvested quantities will be halved by 2050. This will be particularly significant, as the human population will continue its rapid growth.

Consequences of climate change will slow economic growth and poverty reduction. It is estimated that if the climate warms by 2°C, it will cost the world 0.2%–2% of global gross domestic product (GDP) each year. If we do not intervene, the progress that has been accomplished over the past decades may be reversed in the near future, if not worse – its gains may be lost.

It is stated that poor and remote residents of urban and rural areas will be affected the most. Growing poverty and economic shocks will further increase the likelihood of violent conflicts in troubled areas. It is already clear that climate change will more significantly affect developing countries that are based on agriculture and natural resources.

Arctic societies, small islands, and other regions with a high risk of flooding will face new challenges in terms of territorial integrity, shared water resources, and fish stocks. Acid rain and warming waters will create growing problems for coastal populations: humans and living organisms.

The natural environment, the majority of the economy branches, and the human himself are very vulnerable to climate change. At first sight, a temperature higher by a few degrees may seem to be beneficial in temperate climate zones: lower energy consumption, better conditions for recreation, and it is not as "hot as hell". Since 1960, the length of the vegetation period in Europe has grown by about 10 days, and the biomass has increased by 12% during that time. Climate warming of 2°C–3°C with optimal rainfall would raise crop yields by 15%–20% in Northern Europe.

Unfortunately, optimists will be disappointed. Much more problems will arise than might be seen at first glance, and in order to solve them or at least to reduce their impact, more money will be invested than obtained from fuel savings or longer vegetation period. The greatest concern is caused by the rise in ocean levels: during the 20th century, these levels have already risen by 15 cm, and in the 21st century, they can raise by another 20–80 cm. This will lead to new environmental problems, coastal erosion, beach erosion, and groundwater salinisation; moreover, the global tourism routes will change, and coastal populations will have to move further inland (the number of so-called "climate refugees" will increase).

Warming is followed by the growth in the number of weather anomalies, loss of biodiversity, pathogen range expansion, melting of mountain and Arctic glaciers, continuously declining permafrost and seasonal snow areas, thermokarst subsidence, ecosystem change, and rapid desertification processes. A fifth of humanity (about 1.2 billion people) are already lacking clean drinking water. This number may double if warming continues to progress. Climate change increases the likelihood that tropical diseases, such as malaria, will spread to Northern regions, as they will be compatible with climate conditions common to malaria mosquitoes. It is believed that an additional 210 million people will be at risk of malaria if the temperature rises by 2°C.

Increasing global air temperature and rising ocean surface temperature support the development of tropical and temperate cyclones. Every region of the world will encounter specific problems of climate change. For example, in Europe, 64% of all catastrophic events and 79% of economic losses are linked with weather and climate anomalies. In 1981–2000, the number of weather and climate anomalies doubled, and on average annual, damages rose from 5 billion to 11 billion US dollars.

Even though developing countries account only for a small part of GHGs, due to their limited capacity to adapt to climate change, they will suffer from the greatest consequences. GHG emissions from developed and rapidly developing countries (the European Union, the United States, Canada, Russia, Japan, China, and India) exceed 75% of total GHG emissions.

Unlimited burning and use of fossil fuels in manufacturing processes will not only be the root cause of global warming but may also lead to a global energy crisis. If the demand for such fuels continues to increase (according to current trends by 2% annually), for example, the consumption of oil and its products will rise from 84 million barrels per day in 2005 to 168 million barrels per day 30 years later. Only Saudi Arabia and some other Middle East countries are expected to have sufficient reserve capacity.

Landscape, ecosystems, and biodiversity: Eutrophication, desiccation, and disturbance of the balance of ecosystems as well as habitat change, and acceleration of changes in natural succession, i.e., the extinction of some species and the emergence of new species, are expected. The impact on bird migration is already felt, and the species composition of European and Lithuanian brooding birds is undergoing a transformation. Thus, the protected territories will lose some of their value.

As a consequence of the rise in ocean levels and the increase in storm surges, part of the coast of the Baltic Sea and the Curonian Lagoon will be flooded. The rise in water levels will induce more frequent saltwater intrusions into the Curonian Lagoon and a gradual change in the northern part of its ecosystem. Due to the increased frequency of hurricanes and storms, coastal management will require new investments and innovations.

Air quality: The likeliness of temperature inversions (especially at night) will increase, leading to a reduced emission dispersion. The permissible emission limit values will be more frequently exceeded in the cities. With the rise of the air temperature, the amount of harmful tropospheric ozone will also increase. As a result of the expanding duration of the growing season and the rise in the concentration of carbon dioxide in the air, the pollen concentrations and the likelihood of allergic diseases caused by them will also increase.

Air temperature: The air temperature will rise during all seasons of the year. Inter-day air temperature variations will increase. In the 21st century, the average annual temperature may rise up by 1.5°C–5.1°C, with the greatest warming occurring during the winter and summer months. Until 2035, the highest rises in temperature will occur in February, March, and July. In comparison with pre-industrial levels, the increase in global air temperature by 2°C is most likely to take place in the 21st century (around the year 2030).

Air temperature extremes: The number of extremely hot days and warm nights will grow. There will be more heatwaves during which air temperature will be higher. The maximums of annual air temperature will rise more quickly in the south-east and south-west of the country, with the most moderate increase forecast in the west of the country. The recurrence of extremely cold weather will decrease slightly.

Water: A river discharge distribution in time is expected: freshets will be more frequent during all seasons, there will be fewer spring floods, and greater chances of dangerously low discharge in summer. This will increase the biological contamination of surface waters during summer and reduce the overall contamination in spring. More regular recurrences of extreme rainfall will increase the risk of very strong freshets localised in unusual areas. With the rise in air and water temperature, the eutrophication processes will expedite, and the quality of water resources will reduce. The rising sea level and more frequent winter storms will also lead to more regular floods in the Nemunas River Delta.

Soil: As the recurrence of extreme rainfalls grows, water erosion also increases. The shorter periods of seasonal freezing may reduce the period when the soil is resistant to erosion. Higher aridity will lead to lower water reserves in the soil, while wind erosion will intensify. In addition, the structure of the soil will change. It will be impacted by the anticipated more regular drying–rewetting and freezing–thawing processes as well as changes in organic matter content.

Energy: The overall energy costs will decline due to the warming climate. However, the energy costs required for cooling and conditioning will rise. Moreover, hydropower resources will decrease. In particular, river discharge will reduce during the low-water period in summer. The increasing recurrence of short-term extreme summer rainfall will put small hydropower structures at risk. Considerable variations in wind speeds will not allow the capacity of wind turbines to be adequately used. The rise in the number of storms, blizzards, squalls, freezing rain, and hails will damage the overhead power lines. Increasing inter-day winter air temperature changes will produce extremely rapid leaps in energy demand, which may lead to more frequent technical failures in the energy system.

Transport: More frequent storms will disrupt navigation in the port of Klaipėda, increasing losses due to downtime. Frequent droughts and freshets will burden navigation in internal waters and impact the water transport infrastructure. Increased air temperature variations around 0°C in winter and heatwaves in summer will have an adverse effect on roads and railroads. The frequency of meteorological phenomena that are disadvantageous to traffic will also increase. Weather changeability and extreme weather will speed up the corrosion of vehicles. More regular thunderstorms, mists, and runway icing will also lead to more frequent delays at airports.

Industry: Rising air temperature will decrease the energy consumption necessary for the production of one unit at the same time reducing the cost of the product. Nevertheless, increasing temperature extremes will lead to more regular disruptions to industrial activities. During droughts, the activities of companies with high water demand for their production can be stopped. More regular air temperature variations around 0°C may increase frost damage in building structures and atmospheric corrosion of equipment and mechanisms. Climate warming will have a more indirect impact on the industry with increasing GHG emission limits and pollution taxes.

Agriculture: Increase in the vegetation period and growth in active temperatures are expected. The conditions for the growth of thermophytes will be more favourable, and new species of plants will be introduced. The photosynthetic efficiency will improve resulting in a potential 5%–10% increase in crop yields at excess humidity and a likely 15%–20% increase in yields during a year with an optimal humidity level. Rising air temperature will reduce the energy costs for heating farms and other agricultural buildings. On the other hand, the number of agricultural pests and diseases will increase. Overwintering conditions of biennial and perennial plants will change: snow cover will be unstable and last for shorter periods of time, recurrent thaws may lead to soaking and perspiration of crops and spread of fungal diseases, while vegetation renewal may occur when a high probability of frosts still exists. Increasing recurrence of heavy rainfalls and hail will reduce the yield of root vegetables and crops as well as deteriorate the quality of the harvest. The likelihood of prolonged droughts will also grow. Heatwaves and prolonged droughts will impact livestock health, slow their growth, and reduce productivity.

Forestry: Increasing aridity will boost forest fires. The rise in the number of storms and squalls will increase the extent of wind damages in forests. Warmer winters will lead to the arrival of new diseases and pests. The productivity of some tree species will decrease, while the productivity of other trees will increase. The species composition of trees may change (declination in fir and spruce forests, growth in deciduous forests, and the emergence of new species), while the unadjusted soil structure will not be optimal for new species.

Human health: The risk of infectious diseases and epidemics will increase. More frequent heatwaves and greater inter-day variations of weather conditions will be more threatening for people who are sensitive to weather changes; thus, their immune systems will weaken. Deterioration of surface and groundwater quality may lead to an increase in the risk of diseases and epidemics related to the quality of drinking water. The rise in the populations of ticks and blood-sucking insects is expected. The extended vegetation period will lead to the spread of allergic diseases.

As the climate warms, the glaciers melt, and the circulation of water changes. Due to this, very heavy rains will occur in some regions of the world, which will lead to flooding; at the same time, some other regions will be affected by droughts. Also, there is another problem: the water level will rise. According to the forecasts of UN Intergovernmental Group of Experts, in 2100, due to the melting of Antarctic and Greenland glaciers, ocean levels will rise by 18–59 cm, which is a great disaster for some islands, coastal areas, and cities. As a result of the rising sea levels (e.g. in the Netherlands), the amount of cultivated land along the sea and river coasts will be reduced. More regular and intensifying extreme meteorological phenomena greatly increase the risk of crop loss. Climate change can cause serious damage to harvests and have dire consequences for many people around the world. As the climate warms, soil degradation and decreased yields will lead to a decline in food quality, increased number of diseases, and food poisoning. Food security will become progressively more difficult. Variations in temperature and precipitation, which increase the number of extreme situations in agriculture, are already reducing production. Climate change will lead to an increase in the prices of the world's most significant agricultural crops (rice, wheat, maize, and soybeans). Falling rates of agricultural production, rising food prices, and declining consumption will mean an increase in the number of starving people since due to higher food prices people will simply become incapable of buying food products.

In some regions of the world, water and food shortages as well as the devastating effects of natural disasters may provoke conflicts over endangered water and food sources and migration.

Agriculture, forestry, fishery, and tourism are the most sensitive sectors to climate change. Therefore, effective measures to adapt to the effects of climate change should be taken in these areas. Developing countries are the most vulnerable because they significantly depend on economic sectors sensitive to climate change.

1.3 Sources and Dynamics of Greenhouse Gases

In the history of Earth, the climate has changed due to numerous reasons, including changes in Earth's orbit around the Sun, changes in solar energy, and geological changes. Current climate changes differ in the sense that they are caused by human activities that release gas and other pollutants into the atmosphere, alter the energy balance of the planet, and thereby cause climate change. The gas, known as the GHG, "traps" the energy that heats the planet in a slightly similar manner of what is happening in the greenhouse (which is why it is called GHG). This warming is impacted by the increased air and ocean temperatures, drought, melting ice and snow, rising sea levels, changes in rainfall, and floods.

GHGs: The main GHGs are carbon dioxide (CO_2), methane (CH_4), and nitrous oxide (N_2O). Carbon dioxide emissions are primarily emitted by burning fossil fuels such as coal, oil, gas, and peat. In addition, carbon dioxide is released due to poor land management and land-use changes, e.g., deforestation and urbanisation. The agricultural sector, including livestock, manure management, and fertiliser application, is a major source of methane and nitrous oxide emissions.

Over the past 200 years, the main concentrations of GHGs have increased significantly in our atmosphere: carbon dioxide – 40%, methane – 150%, and nitrous oxide – 20%. Atmospheric concentrations of these three GHGs are currently at their highest points in the entire period of concentration measurements from the very beginning.

In the atmosphere, all gases that trap heat can be called GHGs. However, in Annex A of the United Nations Framework Convention on Climate Change Kyoto Protocol, it is determined that GHGs include carbon dioxide (CO_2), methane (CH_4), nitrous oxide (N_2O), HFCs, PFCs, and SF_6. In 2012, nitrogen trifluoride (NF_3) was added to GHGs as a result of the sudden increase in its atmospheric concentrations. The before mentioned gases are called like this because the concentration in the atmosphere increases due to anthropogenic activity (IPCC 2018). According to their different global warming potentials, GHGs are often calculated and presented as carbon dioxide equivalents (CO_2e).

The greenhouse effect is mainly impacted by the interaction of solar energy with GHGs such as carbon dioxide, methane, nitrous oxide, and fluorinated gases in the Earth's atmosphere. The ability of these gases to capture heat causes the greenhouse effect. GHGs consist of three or more atoms. This molecular structure allows these gases to trap heat in the atmosphere and then transfer it to a surface that further warms the Earth (Archer 2007). This continuous cycle results in an overall increase in global temperature.

Figure 1.1 shows the global GHG emissions in 2016. The figure clearly demonstrates that the highest proportion of pollutants is constituted of carbon dioxide (76%).

GHG stays in the atmosphere for different periods of time: from several to thousands of years, it mixes perfectly, and similar concentrations of GHGs are measured in the atmosphere worldwide. The concentration does not depend on where the main sources of pollution are located on the planet. Figure 1.2 shows the evolution of the CO_2 concentration in the atmosphere, which clearly demonstrates the extremely rapid increase in the concentration and the consequent real threat of climate change.

The energy sector accounts for the largest share of carbon dioxide emissions. As Figure 1.3 reveals, global emissions from the energy, transport, and industrial sectors are steadily rising.

Carbon dioxide emissions from energy and industrial production can be recorded due to the combustion of different fuels. The chart below shows the change in GHG emissions from 1900 to 2013 by individual fuel types (gas, liquid, and solid fuels, cement production).

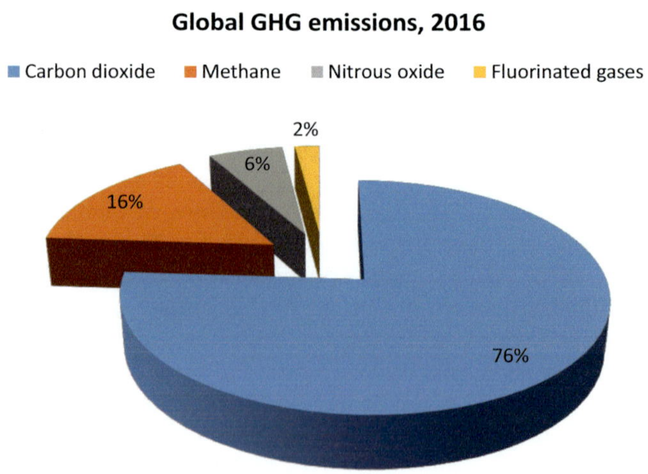

FIGURE 1.1
Global GHG emissions.

Source: Data from https://www.epa.gov/ghgemissions/global-greenhouse-gas-emissions-data.

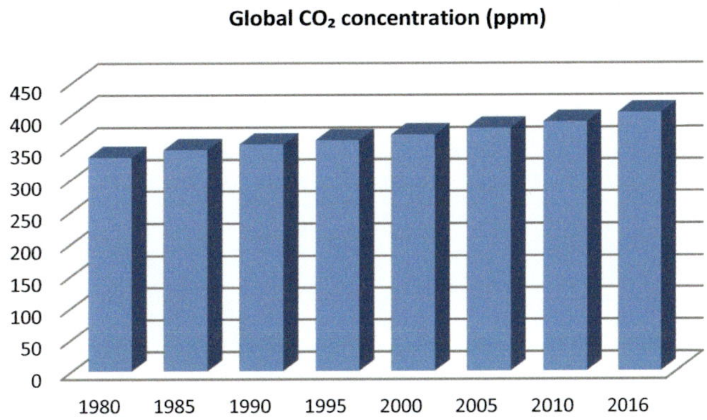

FIGURE 1.2
Global CO_2 concentration.

Source: Global CO_2 concentrations (parts per million) 1980-2016-NOAA.

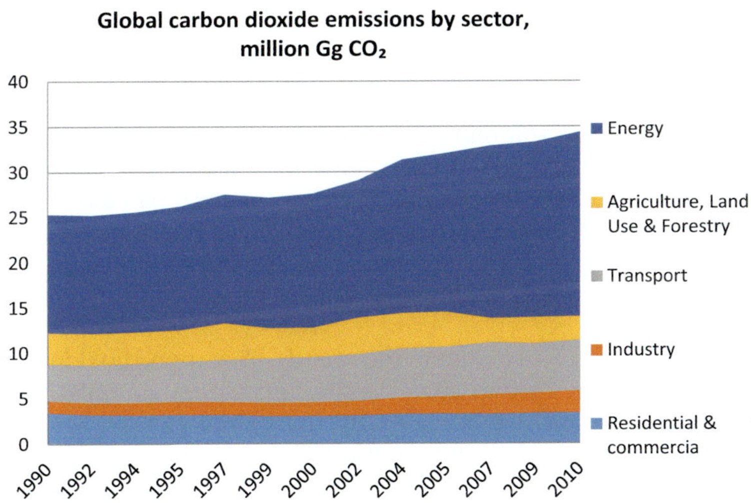

FIGURE 1.3
CO_2 emissions by sectors.

Source: EDGAR (Emissions Database for Global Atmospheric Research) (JRC/PBL) 2016.

As can be seen from Figure 1.4, solid fossil fuels dominate the global CO_2 emission curve during the early industrialisation. Coal-fired power plants first emerged in Europe and North America in the 1700s. Around the late 1800s, reports on pollution from oil and gas burning began. Today, solid and liquid fuels dominate, although gas production contributions are also noticeable. The contribution of each of these sources has changed substantially over time and still shows significant regional variations.

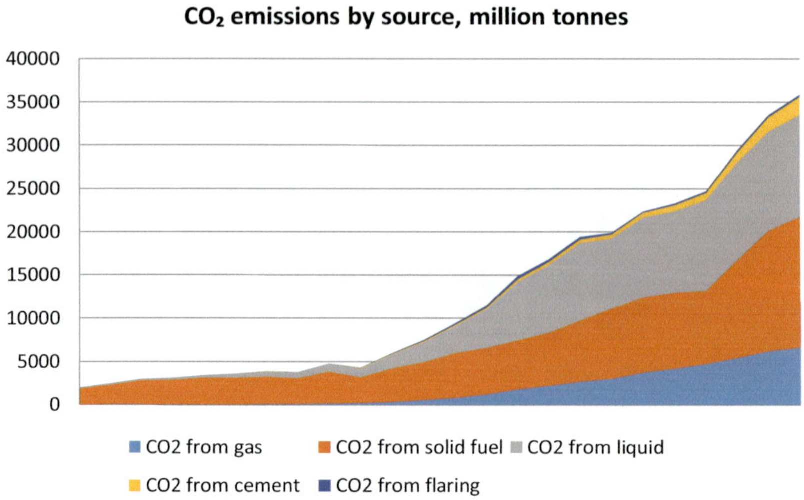

FIGURE 1.4
CO_2 emissions by fuel type.

Source: EDGAR (Emissions Database for Global Atmospheric Research) (JRC/PBL) 2016.

Just 3 years after the stagnation associated with weak economic growth, declining energy intensity, and changes in fuel mixes, in 2017, global CO_2 emissions increased by 2.1%. Strong economic growth has led to increased Chinese coal consumption and CO_2 emissions.

Global economic growth has increased energy consumption and CO_2 emissions in many countries such as India, Russia, Japan, South Korea, Canada, or Iran.

Figure 1.5 provides a chart showing trends in carbon dioxide emissions by world region from 1900 to 2015. As the figure demonstrates, the trends of increasing emissions are revealed by individual countries, the biggest emitters being China, India, and the United States. Annual CO_2 emissions increased from 1.96 Gt in 1900 to 36.18 Gt in 2015.

A list of the most and least polluting countries is compiled and published each year (http://www.globalcarbonatlas.org/en/CO2-emissions). It is important to note that, in these lists, the ranking of countries has been constantly changing since 1960. While the United States has been named the most polluting country in the before mentioned base year 1960, China has been the leading polluting country since 2006, with growth rates (and pollution) rising every year (see Figure 1.6). In 2016, Lithuania ranked 95th out of 216 countries in the world, Latvia – 114th, and Estonia – 90th.

Developed and existing models of climate change indicate that economically strong countries can reap some benefits from climate change. It is worth noting that these countries are furthest from the equator and are usually above sea level. As a result, due to the warmer climate in these countries,

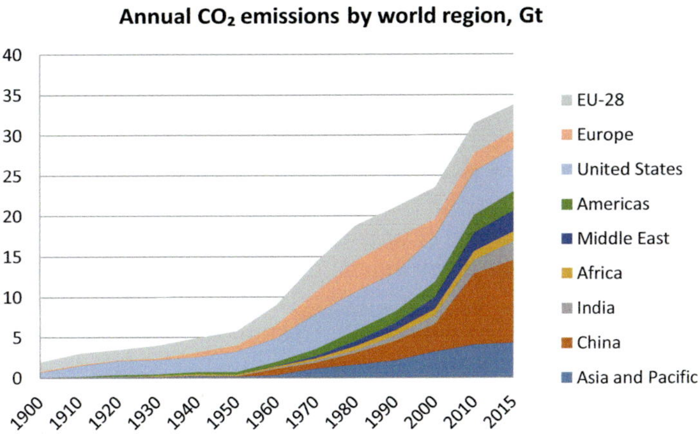

FIGURE 1.5
CO_2 emissions by world region.

Source: Carbon Dioxide Information Analysis Center.

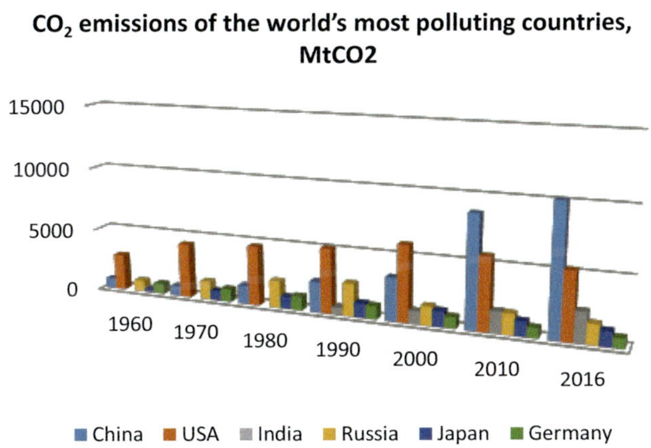

FIGURE 1.6
CO_2 emissions of the world's most polluting countries.

Source: Data from http://www.globalcarbonatlas.org/en/CO2-emissions.

a longer harvest season can be expected as well as facilitated international trade conditions due to shorter trade routes, etc.

The long-held view that climate change and its consequences will be equally important for the wealthy and poor populations of the planet is gradually changing as it becomes clear that no climate change can "equalise" the layers of society. Nowadays and even in extreme conditions, the wealthy can obtain some protection measures that mitigate the situation or at least allow them to survive, while the poor population of the world, caught in the trap of climate change, remains helpless and becomes completely dependent on the poor and increasingly poorly managed grace of the state.

The so-called global justice issue that comes up every time during the climate change negotiations is extremely important. The fact that global warming is caused by so-called GHGs, which have been accumulating in the atmosphere for more than two centuries, has been discovered by scientists and is no longer surprising. Such an indisputable fact presents another fact, which is also difficult to deny: countries, where industrialisation has started earlier, have significantly more harmed the planet than other countries where industrialisation has started later. However, it should be noted that the effects of climate change will be felt first and foremost by those countries that have the lowest GHG emissions. This is due to the fact that these countries are located in a relatively disadvantaged geographical location and are extremely poor. Poverty in these countries not only prevents from addressing the problems caused by climate change but also hampers any development and expansion in the country.

Figure 1.7 provides a chart that clearly demonstrates the enormous difference between economically strong and weak countries in terms of GDP per capita.

In the face of these facts, policy measure makers are beginning to realise that recognising the processes and consequences of climate change is tantamount to losing an ideological battle when the so-called "miraculous" market is no longer able to handle it. The question is whether we need to pass on our cherished values, consumer habits, economic conditions, and general attitudes of climate change to our future generations.

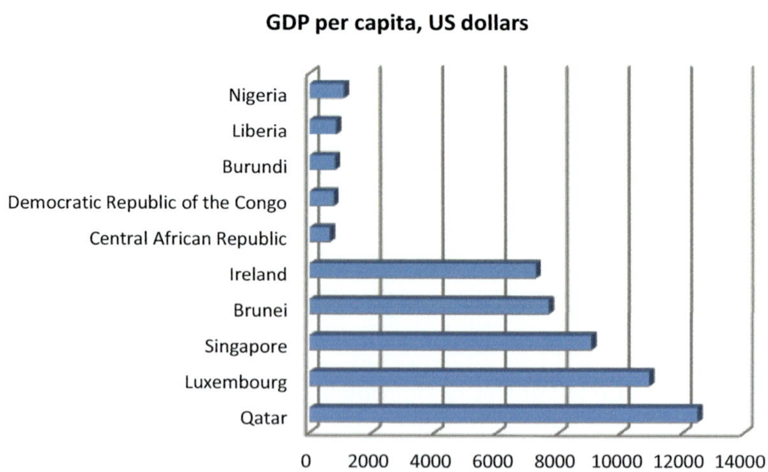

FIGURE 1.7
GDP per capita distribution. GDP, gross domestic product.

Source: Data from https://www.worldatlas.com/economics/.

GHG emission fluctuation trends: GHG emissions and decarbonisation indicators, such as energy consumption, emission intensity, or forest loss, can identify sectors that require urgent action. However, these indicators only describe the status quo and, therefore, have no expected value. The assessment of the climate policy results of each country shows whether there is a policy framework to support the necessary sectoral actions. Ultimately, the transition to technologies with low carbon usage and the green economy is driven by the extent to which financial flows have been directed to a resilient economy with low emissions.

Emerging economies of the Group of Twenty (G20) – in India, Indonesia, and Brazil – have the lowest CO_2 emissions per capita from energy production. As their economies grow, trends change. In accordance with all GHGs and sectors, Indonesia's emissions per capita would be above the G20 average as the agricultural sector emits very large quantities of these gases. For the same reason, Brazil has more GHG emissions per capita than France, Italy, Mexico, and Turkey, respectively.

Mexico and France are the two G20 countries with simultaneously low CO_2 emissions per capita from energy production (2017), which have been steadily decreasing over the last 5 years. In Mexico, the highest CO_2 emissions exist in the energy and heat sector (37%) and transport sector (34%). Recent reductions in CO_2 emissions have been observed in the power generation sector. In France, energy-related emissions are highest in the transport sector (38%), households, and the service and agriculture sectors (26%).

Saudi Arabia and Canada are the two G20 countries with the highest energy-related emissions per capita (2017), and they have not shown emission reductions in the last 5 years. Industry (38%) and energy and heat (38%) account for the largest energy-related CO_2 emissions in Saudi Arabia. In Canada, energy and heat (43%) and transport (28%) account for the largest share of emissions due to high electricity demand and great transport activity.

Different resources and levels of economic development in the G20 lead to different priorities in sectors that require urgent reductions in GHG emissions.

1.3.1 Energy

South Africa (961 g CO_2/kWh), Australia (768 g CO_2/kWh), and Indonesia (755 g CO_2/kWh) have the highest emission intensity in the energy sector, and these countries lack specific actions to phase out the usage of coal. South Africa, Australia, and Indonesia are countries with the highest GHG emissions in the energy sector.

In most G20 countries, electricity and heat production account for the largest share of energy-related CO_2 emissions. South Africa, Australia, and Indonesia are the "brownest" G20 countries in terms of their emission intensity levels of the energy sector in 2017.

South Africa has its highest emissions in the energy sector, which are about twice the G20 average, and has been on an upward trend in recent years (2012–2017). This is due to the strong dependence on organic fossil fuels (coal) and the small share of renewable energy production (4% compared with the G20 average of 24%) (Enerdata 2018). Due to economic development and the need to supply electricity to the poorest people, electricity demand can be expected to increase in the future. Currently, only 84% of its population has access to electricity, and its electricity demand per capita (3,675 kWh/capita) is below the G20 average (3,920 kWh/capita) (World Bank 2018). South Africa has begun to address its high-intensity energy issues in the energy sector; their problem-solving receives an average score for their policy results in promoting renewable energy, taking into account ambitions and policy package.

Based on the project Integrated Resource Plan (IRP) 2018 (not yet formally adopted), South Africa is planning to expand renewable energy from 3.3 GW to more than 25 GW of installed capacity by 2030, equivalent to around 26% of electricity generation, with wind and solar photovoltaic technologies as the main sources of power.

No renewable energy aims until 2050 has been adopted so far. The Renewable Energy Independent Power Producer Procurement Program was suspended in 2016 but resumed in 2018 (Government of South Africa). South Africa also receives an average rating for gradual phasing out of coal. The IRP project foresees the construction of new coal-fired power plants by 2024 but also assumes that by 2050 the coal will be reduced to 20% of its energy supply since coal-fired power plants are expected to operate for at least another 50 years. This adds 1,000 MW to new independent power producers using fossil organic fuels. However, these additional capacities are not required for ensuring energy security; they are not the most cost-effective solution, and they increase emissions (Ireland and Burton 2018).

In *Australia*, the electricity demand per capita is more than twice higher than the G20 average. This is the second-highest emission intensity in the energy sector; however, a recent downward trend has been observed. The country's share of renewable energy in power generation is low (in comparison with the G20) (15% vs 24%) (Enerdata 2018). In addition to the aim to increase the share of renewable energy that will end in 2020 and that is not intended to be replaced, there are no more policies dealing with the reduction of GHG emissions. As a result, Australia is relatively poorly evaluated for its policy of promoting renewable energy.

Despite this federal inactivity, the popularity of renewable energy sources continues to grow due to their economic attractiveness and consumer choice. Australia is also poorly rated due to the gradual phasing out of coal. The Australian government emphasises the importance of coal for energy security, although the local economy supports renewable energy sources. Increasingly more coal power plants are no longer economically viable.

Indonesia has the third highest GHG emission intensity in the energy sector, which has increased in the last 5 years (Enerdata 2018). At the same time, Indonesia has the lowest electricity demand per capita and the second-largest biomass-dependent population after India (IEA 2016). Diversification of energy sources is required to meet the future electricity demand and to fulfil the Paris goals. Results of Indonesia's policy on the promotion of renewable energy sources are seen as poor: it has goals to increase the share of new and renewable energy in the primary energy mix to 31% by 2050.

The government offers tariffs for various renewable energy technologies, but the tariff is based on the average cost of producing electricity (including subsidised energy from fossil fuels), which makes non-subsidised renewable energy projects uneconomic in some regions (Government of Indonesia; Ministry of Energy and Mineral Resources Republic of Indonesia). The country has been given a poor rating for stopping the process of gradual phasing out of coal. The government expects 56 GW of new power to be required over the next decade and plans to cover 21 GW of it by using coal. Cessation of coal use is not being considered (Government of Indonesia).

In the G20, 14 countries need a coal decommissioning plan. Australia, China, Indonesia, Japan, Mexico, Russia, South Korea, Turkey, and the United States are not taking any actions to end the use of coal for electricity and heat generation. Brazil, the European Union, Germany, India, and South Africa have taken some steps but are still lacking a specific plan for decommissioning coal.

Canada, France, Italy, and the United Kingdom have set a date for the decommissioning of coal in line with the Paris Agreement. However, the share of coal in their energy mix is small compared with other G20 countries.

Several G20 countries – Argentina, Brazil, France, Germany, and India – have ambitious aims of new renewable energy and provide strong investment incentives. No G20 country has a target to 100% renew itself by 2050 (Climate Action Tracker 2018).

1.3.2 Transport

France, Japan, and Great Britain have the greatest number of developed plans that can be implemented and seek to reduce emissions in the transport sector. The governments of the United States (5.4 t CO_2/capita), Canada (4.8 t CO_2/capita), and Australia (4.0 t CO_2/capita) have adopted a number of plans and provisions in the transport sector; nevertheless, these countries have the highest transport emissions per capita. The United States and Canada lack proper standards in fuel efficiency, whilst Australia does not have them at all.

The situation is especially worrying in the United States, Canada, and Australia. These countries are the members of the G20 with the highest vehicle emissions per capita, even with increasing trends in the United States and Australia, and with inadequate policies to counteract these trends.

Of all G20 countries, the United States has the highest transport emissions per capita. Moreover, emissions are still rising (Enerdata 2018).

The highest G20 motorisation rate is in the United States (891 vehicles per 1,000 inhabitants). At the same time, in 2017, its share in the electric vehicle

(EV) market was small (1.2%). Norway remains the most advanced EV sales market in the world; in 2017, it accounted for more than 39% of new vehicle sales (IEA 2018). Efforts of the United States are considered meagre because of the failure to gradually phase out fossil-fuel-based light vehicles. However, the country aims that the share of used EVs would reach 50% by 2025 (IEA 2018). In 2018, the United States administration postponed the implementation of fuel efficiency standards that oblige to double the fuel efficiency of new vehicles by 2025.

Canada is the second-highest country for vehicle emissions per capita in the G20, although the emissions have fallen from 2012 to 2017 (Enerdata 2018). The motorisation rate of the country is high: 669 vehicles per 1,000 inhabitants. Compared with the G20 average, 1.1% of the country's EV market share is low (IEA 2018).

The rating of Canada's policy of phasing out low trucks, using fossil fuels, is average. The government has accepted low emission and fuel standards. Some provinces provide financial incentives to buy EVs, although some of them have recently been scrapped; nevertheless, transport fuel taxes, compared with other OECD countries, remain very low. Currently, Canada is preparing a national strategy on zero-emission vehicles and a Clean Fuel Standard (Government of Canada).

Among the G20 countries, Australia is the third in terms of transport emissions per capita. Moreover, the data shows a growing tendency (Enerdata 2018). In Australia, 762 vehicles are registered for every 1,000 inhabitants, and the share of EVs in the country (0.1%) is rather insignificant (Enerdata 2018). The Australian Government provides exemptions from some vehicle taxes for very efficient vehicles and, contrary to other developed countries, does not have any passenger vehicle efficiency or CO_2 emission standards; therefore, passenger vehicles are responsible for the largest share of emissions (Climate Action Tracker 2018).

Emissions from vehicles in several other G20 countries – European Union, France, Germany, Italy, Japan, Mexico, Russia, Saudi Arabia, South Korea, and the United Kingdom – are also above the G20 average. In order to achieve goals of the Paris Agreement, urgent actions need to be taken in each of these countries; only by taking firm actions, the emissions in this sector will be reduced to zero by 2050. The efforts of the transport sector policies of these countries are still weak. Politicians of France, Japan, and the United Kingdom are the only ones with a high rating.

France and the United Kingdom announced a ban on the sale of new petrol and diesel cars by 2040. If this ban was introduced in 2030/2035, it would match with 1.5°C established in the Paris Agreement; despite ambitious goals, France's emissions in the transport sector continue to grow due to the increasing need for mobility and insufficient policies, e.g., to effectively perform multi-modal cargo transportation. Japan aims that the share of EVs would constitute between 50% and 70% of total domestic vehicle sales by 2030. In 2018, the government announced that, by 2050, Japan will only sell electric passenger vehicles and that the amount of total passenger vehicle emissions will decrease by 90% in comparison with 2010 (Government of Japan).

India, Indonesia, and China, which have much lower motorisation rates, are the G20 countries that have the lowest transport emissions per capita. At the same time, because of the economic growth in the last 5 years (2012–2017), the emissions have increased in all three countries; therefore, efficient transport sector policy is vital.

India has the lowest rate of transport emission per capita; however, its 20% rate takes the third-highest spot in the last 5 years (the highest growth rate is recorded in Turkey – 51%) (Enerdata, 2018). The motorisation rate in India is still the lowest with 17 vehicles per 1,000 inhabitants. The market share of EVs in this country remains extremely low (0.06% in 2017) (IEA 2018). For its effort to phase out fossil organic fuel, India gets only an average rating; though the government is encouraging the introduction of EVs, recently it lowered its aim to sell 100% EV by 2030. In 2018, the government announced a new National Electric Mobility Mission Plan, in which the focus is placed on the charging infrastructure and EV procurements by the government (Shah 2018).

Low transport emissions per capita are also recorded in Indonesia, where, over the last 5 years, growth rates have been below the G20 average (1.4%) (Enerdata 2018). In Indonesia, the motorisation rate is 50 vehicles per 1,000 inhabitants; it is higher than the rate of India, but still low in comparison with the G20. From 2018, petrol vehicles must meet "Euro 4" standards, while "Euro 2" standard for diesel vehicles is still valid until 2021 (TransportPolicy.net).

China has the third-lowest vehicle per capita rate in the G20, but its growth rate in the last 5 years takes the second-highest spot (TransportPolicy.net). In China, the motorisation rate is still low with 83 vehicles per 1,000 inhabitants (Enerdata 2018). In 2017, 1 million electric cars were sold worldwide (this is a new record), while half of them were sold in China. This country has the largest market share of EVs (2%) (IEA 2018). China is executing climate change mitigation policy in the transport sector: it has established strict fuel efficiency standards, the highest light vehicle purchases subsidy scheme in the world, and a new "cap-and-trade" system, which obliges the car manufacturers to produce no less than 12% of low emission vehicles by 2020 with the goal to sell 2 million EVs a year (Carbon Brief 2018).

1.3.3 Industry

In the industrial sector, emission intensity is highly impacted by the structure of the industry: many developed countries have relocated their heavy industry companies; therefore, their intensity is lower when emissions are calculated and allocated based on the territorial boundaries. If emissions from goods produced elsewhere were taken into consideration, emissions of developed countries would be higher by 10%–20% (OECD 2018).

Taking into account emissions from industrial production, South Africa, China, and Russia have the highest emission intensity in the industrial sector (Gütschow et al. 2018). In addition, these countries lack ambitious policies with a long-term vision.

In comparison with other G20 countries, South Africa has the highest rate of emissions in this sector. In the last few years, emission intensity has increased only in Brazil and Saudi Arabia. South Africa has implemented some energy consumption efficiency measures in the industrial sector. By applying the Energy Efficiency Strategy of the Republic of South Africa after 2015 (which is not yet approved), the country anticipates reducing energy consumption in the industry, compared with the level of consumption in 2015. Support schemes are mainly constituted by a voluntary energy audit, training, and tax incentive systems; however, they lack the objective to lower the amount of CO_2 produced by new installations with high emission rates (Department of Energy 2018).

China has the second-highest rate of industrial emissions (after South Africa). There is an energy-saving and low-carbon technology programme prepared for the top 10,000 most energy-consuming enterprises in the industry. However, the country does not have a goal that new installations in high emission sectors would produce low levels of carbon; therefore, it lacks a long-term vision on how to reduce emissions in this sector.

Russia is third in terms of emission intensity in its industrial sector, although a slight downward trend of 3% can be noticed. Large energy consumers are required by the country to do mandatory energy audits and transition to the best available technologies by 2025.

None of the G20 countries has established the EU goal to have low-carbon standards applied to new installations in high-emission sectors. The EU Industrial Emissions Directive requires approximately 50,000 industrial installations to obtain a permit which would prove that they are operating in accordance with the best available techniques. Most G20 countries support energy consumption efficiency in industrial production, albeit with varying degrees of ambition. No action is taken in Australia, Saudi Arabia, and the United States.

1.3.4 Forest Sector

From 1990 to 2015, the forest area in Indonesia, Argentina, and Brazil has decreased by 23%, 22%, and 10%, respectively (CAT 2018). Argentina suffers less losses in this respect due to its smaller forest areas.

Indonesia has no national goal to have zero deforestation. Despite the 2011 moratorium on logging in undisturbed areas, which was valid until May of 2019, Indonesia still faces an alarmingly high rate of deforestation, mainly due to the production of cellulose and palm oil. However, the government has suspended the issuance of new palm oil licenses until 2021, suggests reforestation support schemes, and has established an agency for managing the financing of REDD+ activities (Mongabay 2018).

In 2017, Argentina has adopted the National Action Plan on Forests and Climate Change to reduce the GHG emissions in the forest sector by at least 27 $MtCO_2e$ by 2030. Nevertheless, there is no goal to reach zero deforestation.

In 2008, Brazil set a goal to reach zero deforestation by 2015, but the date was later changed to 2030. From 2005 to 2012, the deforestation rate in the Amazon Jungle has decreased by 76%; however, from 2012 to 2017, it increased again by 52%. This latest trend does not meet the objectives of Brazil's NDC. The government plans to replant a 12-million-hectare area by 2030. Moreover, in 2017, it launched an updated monitoring system to fight illegal logging (Timberley 2018; Eisenhammer 2017).

China, Germany, and India seek to achieve at least zero net deforestation by 2020, which would be compatible with the goals of the Paris Agreement.

China aims to increase the forest area of the country from 21.7% to 23% in the period of 2016–2020; therefore, logging must be at zero (Stanway 2018).

According to the 2050 German Climate Action Plan, in the upcoming decades, Germany aims to increase the country's forest area (BMU 2016).

The Indian government is currently reviewing their forest policies in order to meet India's NDC: the aim of the policy project is that at least one-third of the total land would be covered by forests and trees. The current level is at 24.4%; therefore, India aims to increase the total forest area (MOEF 2018).

1.4 Legal Regulation

At the international level, the problem of climate change was brought up for the first time by scientists at the first World Climate Conference in 1979. As public concern had increased, countries began to pay more attention to the discussions about climate change and, in 1988, the United Nations General Assembly adopted a resolution calling for "...protection of global climate for present and future generations of humankind". The United Nations Environment Programme and the World Meteorological Organisation created a new international scientific organisation, the IPCC. This new organisation was commissioned to summarise and evaluate scientific information on climate change. The first climate change assessment report that highlighted the reality of the problem of climate change was published in 1990. The Second World Climate Conference (1990, Geneva) called for the adoption of an international convention. General Assembly of the United Nations has announced negotiations on the Framework Convention on Climate Change. Delegates from more than 150 countries have been negotiating the text of the convention from February 1991 to May 1992.

In the United Nations Conference on Environment and Development (also known as the Rio Earth Summit, Rio de Janeiro), a summit of 178 governments adopted an agenda of the 21st century, which was the global Programme of Action for Sustainable Development and the Declaration on Environment and Development. In June 1992, shortly after the United Nations Conference on Environment and Development, the United Nations Framework Convention on Climate Change (UNFCCC) was signed by 155 countries. The goal of the UNFCCC is to achieve a stabilisation of GHG concentrations in the atmosphere at such level that anthropogenic impacts would not adversely affect the climate system. This level must be reached within a time frame that allows ecosystems to adapt to climate change naturally.

In 1992–1993, 166 countries signed the Convention. The Convention entered into force 90 days after it was ratified by the 50th country, i.e., on 21 March 1994. Ten years later, the Convention was ratified by 188 countries and the European Union.

The main aim of the UNFCCC is to achieve the stabilisation of GHG emission concentration in the atmosphere at such level that dangerous anthropogenic impacts would not disrupt the climate system. This level must be reached within a period which allows ecosystems to adapt to climate changes naturally in order not to endanger food production and the stability of economic development.

The convention lays down basic principles for combating climate change at the international level but does not cover all possible instruments and mechanisms. Therefore, at the first Conference of the Parties (COP) to the Convention, which was held in Berlin in 1995, an agreement was made on more stringent, specific commitments of developed countries, which received the title of the Berlin Mandate.

In 1997, negotiations on the adoption of GHG reduction targets in developed countries culminated in the third meeting of the Parties to the Convention in Kyoto, Japan. The Kyoto Protocol lays down basic rules for reducing GHGs but does not specify their application. The protocol shall enter into force only if it is signed by 55 Parties to the Convention, including Parties of Annex I, which account for more than 55% of the total GHG emissions of the Annex I countries. The Kyoto Protocol entered into force only 7 years later, on 16 February 2005, 90 days after being signed by the Russian Federation. The United States and Australia are the largest emitters that do not support the principles of the Kyoto Protocol – quantitative targets for reducing GHGs.

In 1988, during the fourth session of the Conference of the Parties to the Convention (COP4), negotiations began on the implementation of the Kyoto Protocol, namely on financing and technology transfer issues. The solutions, called Buenos Aires Plan of Action, had to be adopted during the sixth session of the Conference of the Parties (COP6) to the Convention in The Hague, the Netherlands; however, after a failed compromise in solving technical and political issues, the negotiations were extended. Following the exclusion of controversial lines from the Buenos Aires Plan of Action, the Bonn Agreement was signed in July 2001. During the seventh session of the Conference of the Parties to the Convention (COP7), negotiations were concluded reaching a set of detailed decisions (Marrakesh Accords) concerning the Kyoto Protocol and the rules of implementing the Convention. During the eighth, ninth, and tenth sessions of the Conferences of the Parties to the Convention, no substantive political decisions were taken.

The eleventh session of the Conference of the Parties to the Convention (COP11) was historical since it was held in conjunction with the first MOP1 (Meeting of the Parties) which took place at the end of 2005 in Montreal, Canada. During this meeting, the Parties decided to open a dialogue on a long-term policy to reduce GHG emissions. Even without the support of the administration of the United States and Australia, in 2006, 35 Parties that had signed the Kyoto Protocol already considered binding reduction targets after the first round of the Kyoto Protocol in 2008–2012. In addition, it was agreed to revise the Kyoto Protocol. During this meeting, the Marrakesh Accords were approved. These resolutions allow the effective operation of more flexible mechanisms of the Kyoto Protocol (Joint Implementation [JI] projects, Clean Development Mechanisms [CDMs], Emissions Trading System). In addition, COP11 supported the development of a 5-year action plan of adaptation (adjustment) to mitigate the effects of climate warming in developing countries.

The Convention identifies several principles to be followed by the Parties to the Convention, namely:

- The principle of common but differentiated responsibilities: In line with this principle, the Convention states that, first and foremost, developed countries should reverse long-term trends in emissions and take more stringent measures than developing countries. The richest countries should provide financial support and transfer advanced technologies to developing countries in order to help less advanced countries adapt to the adverse effects of climate change. The specific needs and special circumstances of countries in transition need to be carefully studied and taken into account.
- The precautionary principle calls for measures to prevent climate change, including cases when there is no irrefutable scientific evidence of the causes of climate change.
- The principle of the link between climate change and development underlines that energy consumption, land use, and demographic growth are the key elements of both development and climate change. Measures to preserve the climate system need to be integrated into national development programmes.
- The efficiency principle emphasises that climate change policies and measures must be cost-effective and ensuring overall benefits at the lowest possible cost.

Principle of cooperation asserts that Parties should cooperate in the promotion of a stable international economic system, causing the development in all, especially developing, countries in this way enabling them to better address issues associated with climate change.

When selecting measures, it is necessary to take into account the different starting points, economic structure and available resources, efforts to maintain stable and rational economic growth, and available technologies and circumstances that are specific to each country individually. The Convention seeks to ensure that efforts of Parties to reduce and stabilise GHGs are adequate for their emissions and their potential for reduction.

The Parties to the Convention are divided into three groups, which are subject to different requirements:

- Annex I Parties
- Annex II Parties
- Non-Annex I Parties

In Annex I, 41 industrialised countries are mentioned (Table 1.2), including members of the Organisation for Economic Co-operation and Development (OECD) (in 1992), countries with economies in transition, and some countries in Central and Eastern Europe (including the Russian Federation and Belarus).

In Annex I of the Convention, a list of market economies and economies in transition is provided, including the Republic of Lithuania, which shall develop national strategies and apply climate change mitigation measures, by limiting their emissions of GHGs and enhancing the protection of biological GHG sinks (forests). Such strategy and measures would allow countries around the world (the community) to reduce the level of carbon dioxide (CO_2) and other GHGs to 1990 levels by the end of this decade. The Convention covers gases (chemicals) that are not included in the Montreal Protocol on Substances that Deplete the Ozone Layer. When preparing a national climate change strategy, account should be taken of the Parties' unequal starting position, economic structure and raw materials base, efforts to maintain stable and rational economic growth, available technologies, and other individual circumstances; it is also necessary that the contribution of each Party concerning this purpose would be adequate. In accordance with the Articles 4.1 (j) and 12 of the Convention, each Party shall submit a national report of such content through the secretariat of the Convention to the COP, in which the national GHG registry describes the measures the Party has already taken or intends to take to implement its obligations.

Annex II of the Convention includes members of the OECD apart from countries with economies in transition. These Parties undertake to finance the efforts of developing countries to mitigate the effects of climate change

TABLE 1.2

Annex I Parties of UNFCCC

Ireland	Lithuania[a]
Australia	Poland[a]
Austria	Liechtenstein
Belarus[a]	Luxembourg
Belgium	Monaco
Bulgaria[a]	New Zealand
Czech Republic[a]	Norway
Denmark	Netherlands
Great Britain	Portugal
Estonia[a]	France
European Union	Romania[a]
Greece	Russian Federation
United States	Slovakia[a]
Japan	Slovenia[a]
Italy	Finland
Iceland	Sweden
Spain	Switzerland
Canada	Turkey
Croatia	Ukraine[a]
Latvia[a]	Hungary[a]
	Germany

Source: Created by authors.
[a] Countries with economies in transition.

and to adapt to the adverse effects of climate change. In addition, Annex II Parties must promote cleaner technologies for economies in transition and developing countries.

Non-Annex I Parties include developing countries. Some of them are recognised as particularly sensitive to the impact of climate change, either due to low elevation above sea level or due to frequent droughts and severe desertification. The economic development of other countries (e.g. Saudi Arabia) depends on the extraction and sale of fossil fuels and can, therefore, be strongly influenced by climate change mitigation measures. 48 least developed countries have been granted exceptional conditions. The Convention provides for financial assistance and technological development in these countries, helping them to adapt to the adverse effects of climate change.

All Parties to the Convention shall assume their respective general obligations. Every year before April 15, each Party shall provide information to the COP on GHG emissions (national inventory report) and explain the reasons for the change in GHG emissions. The report shall include data on the amount of GHG emissions/removals, taking into account the cyclical nature of the processing and reporting of national statistical data. For instance, in 2006, data for 2004 was provided as well as a GHG inventory summary for all previous years, starting from the base year (1990 or other).

GHG accounting is prepared in accordance with the standardised methodology established by the IPCC, based on national statistical data, and research on country-specific conditions.

Every 3 years, the Annex I Parties shall submit a comprehensive national report on climate change. This is a key document demonstrating the efforts of the country to mitigate climate change by mitigation and adaptation measures, management of sinks, and ongoing research and scientific programmes. According to the national reports on climate change, the COP to the Convention monitors the progress of the Parties and assesses their ability to meet the targets and obligations set out in the Convention.

Non-Annex Parties provide less information, and the detail of the information depends on the amount of funding received for the preparation of the reports. Non-Annex Parties shall submit a national report on climate change within 3 years of the entry into force of the Convention in that country or within 3 years of receiving the necessary funding. Part of the non-Annex Parties – the least developed countries – may submit national reports "at their discretion". The deadlines for the submission of national reports on climate change shall be determined during the Conferences of the Parties to the Convention.

Non-Annex Parties are not required to provide national inventory reports, while experts do not evaluate their national reports on climate change. However, in 2005, the initial national report on climate change, together with the national inventory report, was submitted by 123 non-Annex Parties. Despite significant progress, developing countries are severely lacking in funds and capacity to provide information to the UNFCCC.

The structure and content of the national report are established in the UNFCCC guidelines. Documents drawn up in accordance with these guidelines are more transparent and easily comparable. The Annex I Parties had to prepare a national report on climate change within 6 months of the entry into force of the Convention in that country. These national reports on climate change of Annex I Parties are verified and evaluated by independent experts. Experts shall assess the effectiveness of the means used by the Parties to implement the obligations of the Convention and make recommendations to improve the quality of the report. Since 2003, national inventory reports have been checked annually.

The COP to the Convention is the supreme decision-making body of the Convention which decides on the development, rules, and new obligations of the Convention. The Parties to the Convention shall meet annually, choosing the location on a rotating basis, each time on a different continent. The COP to the Convention is assisted by two subsidiary bodies, which meet twice a year to prepare for the Conferences of the Parties:

- Subsidiary Body for Scientific and Technological Advice (SBSTA) that advises COP on methodological, scientific, and technological issues prepares manuals and methodologies for inventory and national reports.
- Subsidiary Body for Implementation (SBI) that helps to assess the level of implementation of the Convention analyses national reports on climate change provided by the Parties and discusses financial and administrative issues.

The UNFCCC secretariat shall assist the COP to the Convention and its subsidiary bodies. The secretariat organises sessions of subsidiary bodies, collects, analyses, and disseminates information as well as cooperates with other relevant international organisations.

The Global Environment Facility (GEF) manages the financial mechanism of the Convention, which allocates loans and non-repayable subsidies to developing countries. In 1999, the PAF was established by the World Bank, the United Nations Development Programme (UNDP), and the United Nations Environment Programme (UNDP). The PAF supports projects with global environmental impacts on climate change, and the protection of biodiversity, the ozone layer, and international waters. In the field of climate change, the COP to the Convention defines GEF activity guidelines.

The IPCC is the main source of scientific information on climate change. Every 5 years, the panel produces report on the current state of knowledge in the field of climate change. The first report was prepared in 1990, the second report in 1996, the third one in 2001, while the fourth report was prepared in 2006. The IPCC also prepares specific technical reports at the request of the COP to the Convention. In addition, the panel contributes to the preparation of the Guidelines for National Greenhouse Gas Inventories.

The UNFCCC authorities work closely with the so-called Rio Conventions (the UN Convention on Biological Diversity (UNCBD) and the UN Convention to Combat Desertification (UNCCD)). Joint methodologies are developed, joint actions are planned, and potential problems are identified. In addition, the UNFCCC authorities liaise with the World Meteorological Organisation and the authorities of the Ramsar Convention on Wetlands of International Importance especially as Waterflow Habitat.

At the regional level, the UNFCCC secretariat coordinates actions and results with organisations active in the climate change sphere. Active contacts are maintained with national and international non-governmental organisations and associations of business organisations.

The assistance in building capacity on climate change is necessary for the least developed countries, developing countries, and economies in transition. Countries receive support for the preparation of national reports on climate change and for the deployment of climate change mitigation technologies. During the fifth session of the COP in 1999, a Consultative Group of Experts had been set up to analyse the issues of non-Annex I Parties relating to the preparation of national reports on climate change and to put forward proposals to address the current challenges.

The Marrakesh Accords provide capacity building schemes for developing countries and countries with economies in transition. These schemes will help countries to actively engage in the implementation process of the Kyoto Protocol and the Convention. Capacity building must be based on the needs and experiences of the specific country and be in accordance with the learning-by-doing principle and cooperation. The implementation of capacity-building programmes shall be monitored by the SBI.

The introduction of cleaner technologies and the application of sustainable development principles mitigate climate change. The Convention secretariat collects, analyses, and summarises information on technology needs in developing countries and technological development in developed countries. The Marrakesh Accords call for "meaningful and effective actions" in the following areas:

- Assessment of technological needs
- Development of a technological information system
- Establishment of an infrastructure supporting technology transfer
- Capacity building
- Financing of technology transfer schemes

These areas will be funded by the GEF climate change target programmes and the Special Climate Change Fund. The expert group on technology transfer (20 members) supervise the development of technology transfer and report to the Subsidiary Body for Scientific and Technological Advice.

All Parties that have signed the Convention undertake to cooperate in the field of research and observation of the global climate system and to raise public awareness of climate change. Studies and observations are carried out in close cooperation with the secretariat of the Global Climate Observing System. Concerns are raised about the inefficiency of climate observing systems in many regions and the low participation of climate observing teams in developing countries. Information on climate change studies and observations shall be provided in national reports on climate change.

In accordance with the Convention, the Annex I Parties can contribute to reducing GHG emissions for other countries by implementing climate change policies. The first COP meeting launched a "pilot phase of activities implemented jointly". An Annex I Party may implement projects mitigating GHG emissions in the territory of the other party (e.g. energy-saving projects) or increasing GHG removals (e.g. forest planting). Unlike the Kyoto Protocol mechanisms – JI projects and CDMs – JI projects do not create carbon credits.

Through joint projects, the Parties to the Convention were seeking to gain experience in cooperation and project implementation in the field of climate change.

The Kyoto Protocol: The Parties to the Convention are divided into two groups, which are subject to different requirements. In Annex I, 41 industrialised countries are mentioned, including members of the OECD (in 1992), countries with economies in transition, and some countries in Central and Eastern Europe. OECD members (24 countries apart from those with economies in transition) are included in Annex II. Other countries, mainly developing countries (145), are classified as non-Annex I Parties. Lithuania is included among the Annex I Parties. The countries, which are listed in Annex I to the Convention (including the Republic of Lithuania), in accordance with Article 4(2)(a) and 4(2)(b), shall develop a national strategy and take appropriate measures mitigating the consequences of climate change by limiting anthropogenic GHG emissions and improving the protection of the above-mentioned GHG sinks. Such strategy and measures would allow reducing emissions of carbon dioxide (CO_2) and other GHGs, not included in the Montreal Protocol on Substances that Deplete the Ozone Layer, to 1990 levels by the end of this decade. An account should be taken of the Parties' unequal starting position, economic structure and raw materials base, efforts to maintain stable and rational economic growth, available technologies, and other individual circumstances; it is also necessary that the contribution of each Party concerning this purpose would be adequate. In accordance with the Articles 4.1 (j) and 12 of the Convention, each Party shall submit a national report of such content through the secretariat of the Convention to the COP, in which the national GHG registry describes the measures the Party has already taken or intends to take to implement its obligations.

The UNFCCC Kyoto Protocol was adopted at the third session of the COP on 11 December 1997 in Kyoto, Japan. Pursuant to the Kyoto Protocol, during the first commitment period of the Kyoto Protocol (2008–2012), industrialised countries committed to reducing emissions of six GHGs (CO_2, methane, nitrous oxide, HFCs, PFCs, and SF_6) approximately 5% below the 1990 average. A 5-year and not 1-year period was chosen to level out annual fluctuations due to uncontrolled factors such as weather change. No tasks for emission reduction were provided for developing countries.

In the Kyoto Protocol, Parties are grouped into Annexes I and II, while Article 2 of the Kyoto Protocol stipulates that in order to promote sustainable development, the Annex I Parties shall develop and implement policies and measures that would seek

- to enhance energy efficiency in relevant sectors of the national economy;
- to increase the amount of GHG sinks;
- to promote sustainable forms of agriculture;

- to increase the use of renewable energy sources;
- to reduce or phase out market imperfections, subsidies, and tax exemptions in all GHG emitting sectors of the economy;
- to encourage reforms concerning GHG emissions in relevant sectors of the economy;
- to take measures to reduce GHG emissions in the transport sector; and
- to reduce methane emissions from waste management as well as production and distribution of energy.

In addition, the Kyoto Protocol encourages cooperation and exchange of information among the Parties on national measures mitigating climate change.

Article 3 of the Kyoto Protocol states that the Annex I Parties shall, individually or jointly, reduce GHG emissions by at least 5% below 1990 levels during the commitment period 2008–2012 and by 2005 illustrate how this goal is pursued. Articles 3(5) and 3(6) indicate that Parties in the transition to a market economy may, with the approval of the COP, consider another period as their base year.

The Kyoto Protocol aims to reduce the release of emissions of GHGs such as carbon dioxide (CO_2), methane (CH_4), nitrous oxide (N_2O), and chlorine–fluorine–hydrocarbons (hydrofluorocarbons (HFCs), perfluorocarbons (PFCs), and sulphur hexafluoride (SF_6)) into the atmosphere.

Article 3 of the Kyoto Protocol also allows countries to transfer/accept GHG reduction units; thus, it provides prerequisites for international Kyoto Protocol mechanisms. Taking into account national development priorities and local circumstances (Article 10), Parties shall establish, implement, publish, and regularly update national programmes containing mitigation measures. These programmes should cover energy, transport, and industry sectors as well as forestry, agriculture, and waste management. These programmes should be described in national reports. In addition, Parties need to cooperate not only in promoting the development of environmentally friendly technologies but also in conducting research on climate change.

The Parties to the Convention shall establish rules and guidelines for the assessment, reporting, and trading of GHG emissions. Annex B Parties may trade GHG emissions between themselves for the purposes of fulfilling their commitments under the Protocol (Article 17) in addition to other actions taken for climate change mitigation. Following Russia's ratification of the Kyoto Protocol, this document entered into force on 16 February 2005.

The Kyoto Protocol planned to increase the cost-effectiveness of climate change mitigation measures through three flexible mechanisms: JI projects, the CDM, and Emissions Trading. These three investment schemes enable reducing GHG emissions where the introduction of reduction measures is the cheapest. The costs of reducing pollution, due to different levels of technology, vary from country to country, but the environmental impact of reducing GHG emissions will be the same regardless of which country produces lower GHG emissions.

While considering the Kyoto mechanisms, the "right to pollute" has been discussed focusing on the fact that the flexible mechanisms under the Kyoto Protocol do not encourage GHG emission reductions in developed countries; thus, countries can trade bogus credits. The Marrakesh Accords clearly state that the Annex I Parties shall take significant national measures to reduce climate change and that flexible mechanisms can only complement national measures. On the other hand, the Marrakesh Accords do not set quantitative limits on the use of flexible mechanisms.

In order to participate in the flexible mechanisms under the Kyoto Protocol, the Annex I Party to the Convention must have ratified the Kyoto Protocol, comply with the reporting and methodological requirements of the Protocol, and have a national carbon credit registry.

The Kyoto Protocol provides for mechanisms with the help of which Annex I Parties can fulfil their obligations:

- Emissions Trading (an arrangement whereby an entity from one Party buys assigned amount units from entities of another Party).
- JI projects (an arrangement whereby an entity of one Party invests in projects implemented in another Party, in return receiving emission reduction units [ERUs] from the latter).
- CDM (an agreement whereby one Party invests in projects carried out in another Party, which is not included in Annex I Parties, and receiving ERUs in return.

The first commitment period of the Kyoto Protocol ended in 2012. At the Conferences of Parties to the Convention in Durban (South Africa) (2011), and in Doha (Qatar) (2012), Parties to the Convention agreed on a second commitment period of the Kyoto Protocol (2013–2020) as well as the preparation of a new international legally binding agreement by 2015 and its entry into force by 2021.

The Paris Agreement, 30 November–12 December 2015: In the 21st session of the COP to the UNFCCC (COP21), the text of the Agreement on strengthening the global response to the threat of climate change was adopted. This Agreement, approved in Paris, is a legally binding international climate change document that includes commitments from 197 developed and developing countries on GHG emission reductions, adaptation to climate change, and implementation measures. The Agreement entered into force on 4 November 2016; currently, it has already been ratified by 167 countries. At the beginning of June 2017, the US President Donald Trump announced that the United States would withdraw from the Paris Agreement, stating that after leaving the United States would seek to start new negotiations: either by re-entering into the Paris Agreement or by concluding a new agreement under conditions that are consistent with the interests of the United States. Despite the US withdrawal notice, keeping in mind that, according to the Agreement, a withdrawal is only possible 3 years after the Agreement enters into force and 1 year after the withdrawal notice has been submitted, many other Parties to the Paris Agreement have confirmed that the Paris Agreement is appropriate and cannot be renegotiated. The Agreement is ambitious, but no strict rules are set out; it allows each Party to determine how to contribute to the targets combating climate change, which threatens the whole world.

The Paris Agreement applies to the period beyond 2020. Moreover, it replaces the framework provided in the Kyoto Protocol (1997), containing the stipulated commitments of the Parties to the Convention until 2020. Measures and programmes agreed among the Parties in previous agreements, such as the Warsaw International Mechanism for Loss and Damage associated with Climate Change Impacts and the Cancún Adaptation Framework, continue to be applied in the context of the Paris Agreement.

The Paris Agreement sets a long-term target, consistent with the objective of ensuring that global temperature increases are well below 2°C compared with the pre-industrial levels and to seek that temperature increases would remain below 1.5°C in comparison with the pre-industrial levels. Based on the best available scientific knowledge, the Parties have undertaken to reach a balance between human-induced GHG emissions and their removals by sinks in the second half of this century. In order to achieve these targets, the Parties will develop, communicate, and maintain successive nationally determined contributions.

The Paris Agreement foresees that from 2023, on the basis of the latest scientific knowledge and hitherto implementation of data, every 5 years, the Parties will carry out a joint assessment of the progress: the progress will be assessed, and the issues concerning the reduction of emissions, adaptation to climate change, and provisions of the support will be discussed; the successive nationally determined contributions of each Party must be aimed at progress, in comparison with the binding actions in force at that time, and they must reveal the greatest possible level of ambition.

The Paris Agreement sets out a global framework, in order to better implement the UNFCCC and its target to strengthen the global response to climate change threats with regard to sustainable development and efforts to eradicate poverty.

Main highlights of the Paris Agreement:

- a long-term target of keeping the increase in global average temperature below the determined limit in this way ensuring the reduction of risks and impacts of climate change was set;
- an agreement on the commitments and actions of the Parties to reduce their GHG emissions was made;

- it was agreed to inform about the actions and targets of the Parties in order to ensure transparency and observation as well as to set even more ambitious targets;
- it was agreed to increase adaptability to the adverse effects of climate change, to promote resilience to climate change and low GHG development in a way that does not threaten food production;
- it was agreed to ensure that the provided funding was consistent with the downward trajectory of GHG emissions and climate change resilient development.

Attention is also drawn to the provisions of the Paris Agreement, which highlight the target of global adaptation to climate change and the need to address the objectives of its implementation in a successive, continuous manner and at all scales. With the Agreement, the Parties established a global adaptation target: to strengthen the capacity to adapt, increase resilience, and reduce vulnerability to climate change, in order to contribute to sustainable development and to ensure adequate adaptation countermeasures, taking into account the desired temperature target specified in the Agreement.

By the Paris Agreement, the Parties have recognised that adaptation to climate change is a global challenge that needs to be addressed at local, subnational, national, regional, and international dimensions and that adaptation is essential integral part of and contribution to the long-term global response to climate change to protect people, livelihoods, and ecosystems, taking into account the need to immediately respond to the needs of developing countries (Parties to this Agreement) that are particularly vulnerable to adverse effects of climate change.

Climate change adaptation actions should be initiated by countries, orientated towards gender equality, participatory in nature, and completely transparent, taking into consideration vulnerable groups, communities, and ecosystems, based on and guided by the best available scientific knowledge and, as appropriate, traditional knowledge, knowledge of indigenous peoples, and local knowledge systems, with the purpose that adaptation issues would be included in relevant socioeconomic and environmental policies and actions, where appropriate.

In the Paris Agreement, the importance of supporting international cooperation on adaptation is recognised; it is also important to take into account the needs of developing countries, especially of those that are particularly vulnerable to adverse effects of climate change. The Parties should strengthen their cooperation in improving adaptation actions in accordance with the Cancun Adaptation Framework, including

- sharing information, good practices, experiences, and acquired knowledge as well as in cases when it is related to science, planning, policies, and implementation of adaptation actions;

- strengthening institutional structures dedicated to the synthesis of relevant information and knowledge and the provision of technical support and guidance to the Parties, including the measures set out in the Convention for the implementation of this Agreement;
- strengthening scientific knowledge on climate change, including research, systematic observation of climate change, and early warning systems, in a manner that provides useful information to climate services and supports decision-making;
- assisting developing countries in establishing effective adaptation practices, adaptation needs, priorities, the support provided and received for adaptation actions and efforts, and challenges and shortcomings in a manner consistent with encouraging good practices; and
- increasing the effectiveness and longevity of adaptation actions.

Each Party shall, as appropriate, take part in adaptation planning processes and the implementation of actions, including the preparation or development of relevant plans, policies, and/or action papers.

These may include

- adaptation actions, the fulfilment of commitments, and/or ongoing efforts;
- the process of drawing up and implementing national adaptation plans;
- the assessment of climate change impacts and vulnerability that would help to develop nationally determined prioritised actions, taking into account vulnerable people, areas, and ecosystems;
- monitoring, assessment, and collection of appropriate knowledge of adaptation plans, policies, programmes, and actions;
- strengthening the resilience of social, economic, and ecological systems through the means of economic diversification, sustainable management of natural resources, etc.

The Paris Agreement has set out an action plan to limit global warming. It is constituted of the following key elements:

- Long-term goal: Governments have agreed to ensure that global average temperature increase is well below 2°C in comparison with the pre-industrial temperature while making efforts to keep it below 1.5°C.
- Actions: Before and during the Paris Conference, Parties presented comprehensive national climate action plans that aim to reduce their emissions.

- Ambition: Governments have agreed to present their action plans every 5 years, setting even more ambitious targets in each plan.
- Transparency: Parties have also agreed to inform each other and the public of their progress in achieving their goals to ensure transparency and observation.
- Solidarity: The European Union and other developed countries will continue to provide financing to developing countries in order to fight climate change by helping them to reduce emissions and increase their resilience to the effects of climate change.

Despite all the previous concerns, the outcome of the Paris Conference is a comprehensive and detailed document, while its ambitious targets and detail are positively assessed. The latter was needed in order for the provisions of this document to be put into practice. Even before the Paris Conference, there have been some discussions that this agreement will be described in history textbooks. And indeed, this is the first agreement of such magnitude in human history where practically the whole world has agreed to reduce climate change by making binding commitments. For the first time, human rights are mentioned in the preamble of the Agreement. It took two decades to reach this Agreement. This Agreement is expected to enter into force in the next year when it will be ratified by at least 55 countries whose combined GHG emissions account for at least 55% of global emissions.

The Paris Agreement is a clear signal of the intention to phase out fossil fuels. This is how the intention to bring GHG emissions closer to zero by the middle of this century, i.e. in the next three decades, should be understood. The deal to reduce GHG emissions is equally binding on all Parties. However, in the Agreement, a difference exists between developed and developing countries: developed countries are obliged to take rapid and effective measures to transition to low-carbon technologies, whereas developing countries commit to actively contribute to the atmospheric pollution reduction in accordance with their capabilities and support from developed countries.

Prior to the Paris Conference, the Parties had already drawn up the so-called Intended Nationally Determined Contributions that were national plans on the reductions of GHG emissions. However, these plans were too weak, and the measures envisaged in them could have led to an increase in the global average temperature even by 3°C. Thus, it has been achieved that the Paris Agreement would initiate a revision and strengthen the plans every 5 years, starting from 2018. The plans will be reviewed in light of the overall global situation and evaluated on the basis of the latest scientific knowledge.

The Agreement also covers adaptation to climate change: Parties shall develop adaptation plans while developing countries shall receive financial, technological, and capacity-building support to implement these plans.

The Agreement also makes a clear distinction between adaptation to climate change and the mitigation of adverse effects of climate change. In order to address the latter issues, a targeted programme to address loss and damage associated with climate change impacts will be established.

According to *Carbon Brief* estimates, in order for the developing countries to be able to implement currently presented national actions by 2030, they need 3,500 billion US dollars. Financing climate change is therefore undoubtedly one of the most important aspects of the Agreement. The Agreement states that developed countries will continue to be the leaders in mobilising climate change finance. The fact that the Agreement separates funding for climate change mitigation, adaptation as well as loss and damage programmes is a positive thing. However, the Agreement left the long-term financial support to developing countries undefined, with neither the amounts nor more specific principles specified on how Parties will deal with this issue in the future. Developed countries will voluntarily contribute to climate change financing and publish their contributions every 2 years. The Agreement still maintains the provision confirmed in the Copenhagen Agreement that, starting from 2020, every year developed countries will spend 100 billion US dollars on climate change mitigation and adaptation each as well as will allocate a larger share than hitherto for financing adaptation processes. Therefore, in the coming years, one of the biggest challenges and tasks for Parties will be to agree on how to ensure this funding for achieving the ambitious goal of stopping climate change at 1.5°C.

The Agreement did not leave the GHG market on the side-lines. The Agreement indicates that climate change mitigation efforts can be transferred to other countries and provide for the possibility of linking the GHG trading markets. It is significant to mention that the Agreement indicates the introduction of restrictions on international trade; thus, Parties shall step up their efforts to reduce their domestic GHG emissions. As a result, in the coming years, rules will be developed on how GHG reductions shall be accounted and how to avoid double accounting of GHG emissions. The Agreement also provides for the establishment of two non-market trading mechanisms. One will focus on reducing GHG emissions and promoting sustainable development, while the other will promote "integrated, holistic, and balanced non-market approaches".

Needless to say, that, the same as other agreements, the Paris Agreement has its flaws and weaknesses, and some parts are missing from it. International aviation is responsible for 5%, while shipping accounts for 3% of global GHG emissions. These two sectors together produce such amounts of emissions that, if they were a country, it would be among the top ten of the biggest emitters. However, these sectors are regulated not by a specific country, but by separate United Nations organisations: The International Civil Aviation Organisation and the International Maritime Organisation. The Paris Agreement is criticised for not highlighting the impact of the two sectors, as current trends are frightening: both sectors are growing rapidly, but the energy and fuel efficiency in them are not increasing. On the contrary, the *Transport & Environment* study shows that over two decades from 1990 to 2010, the fuel efficiency decreased by 10% in the shipping sector. Without intense measures, it is projected that by 2050, these sectors can account for up to 39% of global GHG emissions. However, the situation is not hopeless: The Agreement calls for all sectors to contribute to the maintaining of the global warming below 1.5°C, and it is expected that during the next year, both of the UN organisations will take actions encouraging sectors to increase energy efficiency, change old technologies to more effective ones, reduce ship speeds, use more renewable fuel and, in principle, review and change their activities.

The Agreement is also criticised for not mentioning the agricultural sector separately, and for not taking into account the fairness of climate change: it is apparent that the countries that are the most affected by climate change have historically contributed the least to its growth.

In Article 14 of the Paris Agreement, it is established to aim for the agreed goals by assessing the collective progress

1. to maintain the global average temperature increase to below 2°C and continue efforts to limit the increase to 1.5°C;
2. to increase adaptability to the adverse effects of climate change and resilience to changing climate conditions; and
3. to align all financial flows with the pathway towards low GHG emissions and climate-resilient development.

In order to comply with the Paris Agreement and the determined temperature limit, the CO_2 emissions should be significantly reduced. The greatest progress must be achieved around 2020, while, in 2050, CO_2 emissions should fall to net zero.

According to the 2015 Paris Agreement, the European Union committed to reduce GHG emissions, such as CO_2 and methane, by 40% in comparison with the 1990 levels in all sectors of the economy by 2030.

The EU Member States and the European Parliament (EP) have approved an ambitious project which, according to Brussels, will help the block to meet its climate change commitments under the Paris Agreement. The European Union has decided to reduce CO_2 emissions from new light vehicles by an average of 37.5% by 2030 in comparison with 2021 indicators. Emissions from new vans will have to be reduced by 31%. By adopting these ambitious targets, Europe is once again showing how to turn the Paris Agreement and COP24 (Katowice Climate Change Conference) into specific actions.

2

Sustainable Development and Climate Change

2.1 The Concept and Objectives of Sustainable Development

By analysing the historical development of the concept of sustainable development, the stages of the historical development of the relationship between humanity and the natural environment are often distinguished. The first stage covers the period before the Industrial Revolution and is often treated as a period of human life in harmony with nature by scientists. In fact, at this stage in the evolution of the relationship between the individual and nature, nature dictated its conditions to the human, while the human became awed by nature, often deified it, and tried to accommodate to it as much as possible. In this case, Pagan traditions – sacred groves, sacred rivers, and lakes – well reflect the essence of this approach.

The second stage in the historical dimension of human–nature relations is associated with rapid advances in science and technology, which have led to the formation of a worldview of the conquest of nature, focusing on the fact that humans can conquer nature by changing it according to their own needs and discretion. Constantly accelerating economic development and increasing production volumes have enhanced the growth of the population and increased demand for natural resources, while the increasing use of natural resources has enlarged environmental pollution, which from local or regional has grown to the global level. At this stage, the environment is seen by humans as the building material for creating a more perfect, better world; thus, the humanistic paradigm is displaced by the technocratic paradigm. The core of this stage is the formation of a long-lasting confrontation between the human and nature, which has allowed humans to feel independent from the surrounding nature.

The third stage, explaining the relationship between the human and nature, is often referred to as the stage of human–nature harmonisation with the main aim of reducing the negative effects of economic development on the environment. It emerged by questioning whether the humankind is indeed on the path to full advancement, whether the social and natural environment will be continually improved, and whether material prosperity, regardless of exploitable natural resources, will be achieved without harming the society and the surrounding environment. Precisely in the third stage, environmental problems have been viewed not as negative changes in nature, but as social problems caused by people in their daily lives (Čiegis 2004). In order to solve these problems, it is necessary to use not only the latest tools of scientific and technical progress, but also the inclinations of society, their culture of consumption, their habits, and attitudes established over time.

As a result of these historical stages of the relationship between humanity and the natural environment, the necessity of assessing the responsibility of humanity for their actions arose, as it became evident that further uncontrolled expansion of the technosphere threatens to completely destroy the vital elements. At the same time, not the question "Can the biosphere adapt to change?" but another one arose: "Can the biosphere adapt to change so quickly?" The question of speed arose because the pace of human development surpassed all expectations. At the same time, a new approach to economic growth formed and the understanding and meaning of these concepts began to change, as it became clear that the growth process could be destructive (Čiegis 2004, 2006).

Many postulates on harmony development between humans and environment, in particular the living part of it (flora, fauna), are also easy to find in various religions, especially Buddhism, where a human is considered an integral part of nature. On the other hand, we have to admit that Christianity, which has elevated the human as the perfect and God-like creature above the living nature, is far less favourable to the harmonious relationship between humans and nature. Thus, it is no coincidence that as the civilisation evolves, first and foremost, the deep problems of human–environment relations emerge namely in Christian countries.

The discussions on the relationship between economic growth and the environment are long-standing, complex, and fundamentally unresolvable to this day. The origins of these disputes, though linked to Adam Smith's work An Inquiry into the Nature and Causes of the Wealth of Nations, published in 1776, are often traced back to the ideas of Thomas Robert Malthus outlined in his famous work An Essay on the Principle of Population As It Affects the Future Improvement of Society. The ideas of Malthus were in stark contrast with the ideas about the need to ensure a continuous process of economic growth expressed by Adam Smith in The Wealth of Nations. Thomas Robert Malthus, who developed his own population theory, stating that population dynamics are limited by declining efficiency of the resources, claimed that the population grows in geometric progression, while means of subsistence only in arithmetic progression at best. Thomas Malthus was right in speculating that the population will grow geometrically but was wrong in claiming that food production will increase arithmetically, because he did not foresee humanity's ability to geometrically improve the technology used in agriculture, and did not realise that the economic law of diminishing results is valid only under the immutable status of technologies. Consequently, although Thomas Malthus presented a strong model, its simplicity limited its relevance to political decision-making (Čiegis 2004).

It is worth noting that the problems (overpopulation) raised by Thomas Malthus, Karl Marx's teachings on misallocation, and John Maynard Keynes's doctrine of obligatory labour, all had one common solution to their problems – economic growth. Growth was the general response to all problems. Thus, this expansionist view of the world – a panacea for unmanageable and unbridled economic growth – is also one of the threats to the development of the modern world. Talks over environmentally balanced development to eliminate and prevent (current) consequences of unrestricted economic growth began only in 1983.

The modern version of the neo-Malthus theory was studied by Ehrlich (1989) and Hardin (1968, 1993). They extended pessimistic ideas and stated that uncontrolled population growth would undermine life-sustaining functions and lead to environmental, social, and economic disasters. The identification of a set of such problems can be seen as the beginning of the formalisation of the sustainable development process.

Subsequent ideas focusing on the interpretation of the optimal relationship between humanity and resources were explored in Daly's works Toward a Steady-State Economy (Daly 1980) and Steady-State Economics (Daly 1991). Other classical economists, such as David Ricardo and James Mill, also understood the limitations of natural resources available to humanity and their impact on the scale of economic activity and, therefore, considered long–term economic growth difficult to achieve. Land constraints caused the main concern for resource scarcity. David Ricardo predicted that with the growth of population, land shortages would increase and society would be forced to cultivate increasingly poorer land and, as a result, the price of agricultural products would rise. However, David Ricardo believed that human resources and technological development would help to delay the period when humanity's needs would exceed the possibilities of natural resources (Čiegis 2006).

For a long time, in the discussions about the relationship between economic growth and the environment, a pessimistic attitude prevailed, which is best reflected in the famous report The Limits to Growth written to the Club of Rome by Donella Meadows et al. (1972). The report aimed to clarify the limits of production expansion and population growth, while the starting point was the frightening exponential growth of the population and the economy. The conclusions reached by the scientists were disappointing: the contradictions between nature's limitations and the extremely rapid growth pace of its use, the increasing environmental pollution and the rapid population growth in the mid-21st century could lead to a global ecological crisis. The Limits to Growth and the subsequent projects developed on the initiative of the Club of Rome, such as Beyond the Age of Waste, Goals for Mankind, formed a new mentality and brought the need to ensure global equilibrium to the forefront (Čiegis 2006).

Around 1935, when, through the millennia, the mankind had already been well adapted to the local ecological systems, the English scientist Arthur Tansley introduced the concept of "ecosystem" as a scientific category, which was defined as a dynamic complex of living organisms (plants, animals, and micro-organisms) and non-living environment (soil, water, and air). While the scale of humanity's economic growth was relatively small, it was possible to ignore the fact that the human economy was involved and dependent on the ecosystems of our planet (Čiegis 2004). However, in the long run, there was no area of nature left untouched by a human and without a proper assessment of the ecosystem's condition for a long time, humankind faced ecological problems.

In 1969, at the session of the UN Council, global crisis was brought up since, through continued economic growth and the waste of natural resources at excessively fast rates, it equally threatened both developed and developing countries.

In 1972, at the United Nations Conference on the Human Environment in Stockholm, the connection between economic development and environmental impact was recognised, and the term "ecological development" was proposed. In the Declaration, adopted during the Conference, the key principles were expressed, which must be followed by states in rationally harmonising the relations between development and environment. The first principle of the Declaration proclaimed that "Man has the fundamental right to freedom, equality and adequate conditions of life". The other articles of the document outlined the policy principles and guidelines on rational use and protection of renewable and non-renewable natural resources, relations of economic development and environment as well as scientific and technical progress. The Conference also developed a comprehensive plan of measures with 109 recommendations that covered the most pressing questions of the environmental benefits of that period (Report of the United Nations Conference on the Human Environment 1972).

This concept of global resource management gave Western countries hope for a comprehensive solution to the problems of environmental pollution. However, developing countries made it clear that their objectives and development policies (or, more specifically, economic growth) of that period were a much higher priority than environmental concerns. And yet, notwithstanding all the contradictions that had arisen among countries of different development levels, for the first time, environmental issues were reviewed in a global context, and the idea of the necessity to link ecological problems and their solution to economic and social development was expressed at the Stockholm Conference (Čiegis 2006).

The origins of the current concept of sustainable development date back to 1980, when a vital document – World Conservation Strategy – was published on behalf of the International Union for Conservation of Nature (IUCN), the United Nations Environment Programme, and the World Wildlife Fund. This document completely abandoned the opposition between nature conservation and economic development, postulated in the concept of human survival, and explicitly declared that development and protection were not contradictory, while the rational use of natural resources was an integral part of social development and nature protection. This was the path towards the widely recognised concept of sustainable development (World Conservation Strategy 1980).

As the concept of sustainability was being developed, a significant role was played by the 1986 Ottawa Conference on Conservation and Development organised by IUCN and held to assess the World Conservation Strategy and set out guidelines for its review; moreover, justice and social legitimacy were considered such significant aspects that even the phrase "sustainable and equitable development" was used in the Conference. In the documents of the Conference, the need for a radical change of the old development model is indicated, noting that we need an alternative society, another type of development associated with structural change (Čiegis 2006).

In 1987, the sustainable development philosophy was formalised with the report Our Common Future by the UN World Commission on Environment and Development that had an infinitely noble goal to make the world prosper and be generous to all (Our Common Future 1987). In this report, for the first time in the history of the evolution of the concept of sustainable development, a qualitatively new notion of economic growth has been defined: it is rapid and, at the same time, socially and environmentally sustainable (responsible) economic growth. From this point on, it was agreed that sustainable development encompasses three dimensions – economic, social, and environmental – and that all economic growth shall be achieved by considering the remaining two aspects of economic growth (social and environmental).

Thus far, sustainable development issues have been fundamental and insignificant in the international climate change policy: they have been fundamental in the sense that the reduction of greenhouse gas (GHG) emissions and economic development have often been seen as contradictory. Over the years, policymakers have repeatedly expressed their concerns that ambitious climate policies are limiting development, reducing the number of jobs, damaging the industry, and lowering the standard of living.

At the same time, the link between climate change and sustainable development has hardly been studied. In climate policy, the risk of "carbon leakage" is discussed in detail: this ambitious climate policy can stimulate the relocation of industrial production to other countries; however, the discussions have traditionally focused on calculating GHG emissions and discounting the wider social and economic impacts of climate change actions such as benefits and costs (Ürge-Vorsatz et al. 2014).

Policymakers often do not take climate change into account when planning their economic development. This is clearly illustrated by the current state of the energy sector where coal combustion is still used, and the state of the transport sector, which is reflected by the transport infrastructure that does not consider climate change. This is due to the fact that different ministries have different approaches and due to a lack of policy coherence. The situation is expected to change and improve significantly with the implementation of two milestones: The Paris Agreement and the 2030 Agenda.

The Paris Agreement is the first global agreement on sustainable development based on data assessment and analysis. The legally binding provisions for information flows are aimed at a civilisational transformation and transition of individuals from rural poverty to the urban middle class, while maintaining a global level of well-being and not crossing ecological boundaries.

Key elements of the Paris Agreement:

- Long-term target: Governments have agreed to ensure that the global average temperature increase is well below 2°C compared to pre-industrial temperature while making efforts to keep it below 1.5°C.
- Actions: Before and during the Paris Conference, countries presented comprehensive national climate action plans that aim to reduce their emissions.
- Ambition: Governments have agreed to report on their actions every 5 years, in this way setting even more ambitious targets.
- Transparency: They also agreed to inform each other and the public of their progress in achieving their goals to ensure transparency and observation.
- Solidarity: The European Union and other developed countries will continue to provide financing to developing countries in order to fight climate change by helping them to reduce emissions and increase their resilience to the effects of climate change.

In 2015, the General Assembly of the United Nations adopted the resolution Transforming our World: the 2030 Agenda for Sustainable Development.

2.1.1 Sustainable Development Goals

The resolution Transforming our World: The 2030 Agenda for Sustainable Development signed by the President of the Republic of Lithuania and heads of other 192 countries at the General Assembly of the United Nations officially came into force at the beginning of 2016, introducing Sustainable Development Goals to governmental authorities. This resolution is the result of thorough and comprehensive 3-year negotiations that involved international, national, and regional players from intergovernmental, governmental, and regional institutions, public and private sectors, and the civil society. These goals are important at both global and national levels and encompass both global and national actions with the aim of making our planet more sustainable. The 2030 Agenda is much more ambitious than the Millennium Development Goals and covers a bigger number of issues, and it shall be implemented by developing and developed countries. In the Agenda, 17 Sustainable Development Goals, based on the achievements of the Millennium Development Goals, and including new ones – climate change, economic inequality, innovations, sustainable consumption, peace and justice as well as other priorities – are established. In order to accomplish these goals, a system of 99 indicators has been established, which may fundamentally differ from the system of indicators for assessing sustainable development formed by the United Nations.

The Sustainable Development Goals set out in the Agenda:

Goal 1: End poverty in all its forms everywhere. This goal is ambitious, but it is believed to be possible. In 2000, the world had committed to half the number of people living in absolute poverty by 2015, and this goal was achieved. However, more than 800 million people around the world still live on less than USD 1.25 a day – that corresponds to the total number of Europeans living in extreme poverty. The time to completely eradicate all forms of poverty is now.

Goal 2: End hunger, ensure food security and improved nutrition, promote sustainable agriculture. Malnutrition causes almost half (45%) of deaths of children before reaching age five; that is up to 3.1 million children each year. Across the developing world, 66 million primary school-age children attend classes hungry with 23 million of them being in Africa alone. However, in the past 20 years, hunger has dropped by almost half. Many countries that have suffered hunger and poverty in the past are able to satisfy the nutritional needs of the most vulnerable people. This is an incredible achievement. The time to end poverty and hunger as well as to take care of a healthy diet of the population is now. This requires the support for the development of sustainable agriculture and promotion of small and medium farmers, who supply high-quality food for the locals.

Goal 3: Ensure healthy lives and promote well-being for all at all ages. Everybody knows that the most important thing in the life of all individuals is their health. It has the greatest impact on our quality of life and on what we create and what we can create. Therefore, the realisation of this goal will ensure that every individual has access to the healthcare system of good quality, effective medication, and vaccines. By the time the already mentioned goal was set, children and maternal mortality rate had plunged by more than half in 25 years, which means that this goal can be achieved. However, some problems still need to be tackled such as AIDS, which is the leading cause of death among adolescents in sub-Saharan Africa.

Goal 4: Ensure inclusive and equitable quality education and promote lifelong learning opportunities for all. Poverty, armed conflicts, and other extreme situations have forced many children around the world to stop going to school. There are many regions around the globe where children do not even have access to primary education and many vulnerable households where children do not have the opportunity to go to school. However, since the year 2000, the significant progress towards primary education for all children worldwide has been made: the overall education level in developing regions has reached 91%.

Goal 5: Achieve gender equality and empower all women and girls. The progress that the world has made in the area of gender equality is commendable, but not all problems have been solved. Girls and women still face discrimination not only in public life but also in the labour market. Significant pay gaps still exist today as well as many unpaid "female jobs", e.g., childcare and work at home. The examples of women discrimination in public decision-making and political participation have also been observed.

Goal 6: Ensure availability and sustainable management of water and sanitation for all. Every individual on Earth should have access to drinking water. This is the goal of the year 2030. Although many people take clean drinking water and sanitation for granted, it should be kept in mind that water scarcity directly affects more than 40% of the world's population. To put it in other words, more than 3 billion people suffer from the lack of drinking water and unsanitary conditions in which they have to live. It is believed that this number is going to soar on account of climate change. If we continue along this path we are on now and not make any changes, by the year 2050 at least one in four people will experience the shortage of drinking water. The new path is more focused on international cooperation in protecting swamps and rivers as well as sharing water treatment technologies that lead to the goal accomplishment.

Goal 7: Ensure access to affordable, reliable, sustainable, and modern energy for all. From 1990 to 2010, the number of people with access to electricity increased by 1.7 billion. This is a gratifying achievement. However, as the world's population continues to grow, more and more people will need cheap electricity to light up their homes and streets, use their phones and computers, and work every day. We generate energy by burning fossil fuels and, thus, increasing GHG emissions, which has a direct impact on climate change and causes many problems in every part of the world. Hence, we have to become a more energy-efficient society and make larger investments into more clean energy sources, e.g. solar or wind energy. By these means, we will meet electricity demand and protect the environment.

Goal 8: Promote sustained, inclusive, and sustainable economic growth, full and productive employment, and decent work for all. A significant part of economic growth is a sufficient salary, which ensures the quality of life. The good news is that the middle class is growing all over the world: it has almost tripled in developing countries over the past 25 years. Today, however, job growth is lagging behind the growing workforce. Therefore, entrepreneurship must be encouraged to ensure new workplaces by eliminating forced labour, child exploitation, and human trafficking.

Goal 9: Build resilient infrastructure, promote inclusive and sustainable industrialisation, and foster innovation. Technological progress helps to deal with great global challenges, e.g. create jobs and use electricity more efficiently. For instance, the world is becoming more and more interrelated and thriving because of the common interest that results from the globalisation processes. Opportunities to take advantage of knowledge management, wisdom, and benefits of the virtual world are increasing. However, 4 billion people, most of whom are from developing countries, do not have access to the Internet. Narrowing the digital divide, promoting sustainable industrial development, and investing in research and innovation are important ways to promote sustainable development.

Goal 10: Reduce inequality within and among countries. We must approve the policy that creates an opportunity for everyone, regardless of who they are or where they come from. Income inequality is a global issue that requires global solutions. This implies that the following steps need to be taken: improving regulation of financial markets and institutions, sending official development assistance where it is most needed, and helping people to migrate safely so that they can take advantage of the opportunities of the global world. Together we can now change the direction of inequality from the old history.

Goal 11: Make cities and human settlements inclusive, safe, resilient, and sustainable. Currently, more than half of the world's population does live in cities, and this number will account for about two-thirds of humanity by the year 2050. Cities are getting bigger. In 1990, ten "big cities" with a population of 10 million or more existed in the world; in 2014, 28 million cities with 453 million people were counted. Many people prefer living in cities due to their cultural centres that exist there; however, at the same time, they are gradually becoming centres of poverty. In order to make cities sustainable for all, we can create good, affordable public housing. We can invest in public transport, create green spaces, and get a wider range of people involved in urban planning. By doing so, we can keep things that we like as they are and change the ones that we disfavour or that are just harmful.

Goal 12: Ensure sustainable consumption and production patterns. The level of consumption of some people is particularly high while others consume very little. There is a considerable difference or so-called consumption inequality that manifests itself in a very small proportion of the population, whose consumption is particularly high, and in the majority of the population that consume very little, often even too little to meet their essential needs. Our aspiration should be the world in which everyone has access to the things they need in order to survive and prosper. The consumption level should be maintained in a way that preserves natural resources so that our children would be able to use them as well as their children. The most onerous part is how to achieve this goal. We can manage our natural resources more efficiently and handle waste correctly.

Goal 13: Take urgent action to combat climate change and its impacts. Every country in the world is facing drastic consequences of climate change. Direct loss due to earthquakes, tsunamis, tropical cyclones, and floods is estimated at hundreds of billions of dollars per year. We can reduce the damage of property and losses of life by helping more vulnerable regions, e.g. inland countries and island countries which have to become more resistant to the consequences of climate change. By taking into account political will and technological measures, it is still possible to limit the increase of average temperature on a global scale to 2°C compared with the pre-industrial level. Thus, the catastrophic consequences of climate change may be avoided.

Goal 14: Conserve and sustainably use the oceans, seas, and marine resources for sustainable development. Ocean temperatures, chemistry, and their life forms are incredibly important components of our planet. Survival of over 3 billion people depends on the diversity of seas and coasts. Oceans absorb about 30% of carbon dioxide emitted by humans during manufacturing processes, but emissions have been going up recently, and consequently, oceans become more acidic – since the beginning of the Industrial Revolution, the acidity has increased by 26%. According to the latest estimates, on average 13,000 pieces of plastic litter float on every square kilometre of the ocean.

Goal 15: Protect, restore, and promote sustainable use of terrestrial ecosystems, sustainably manage forests, combat desertification, and halt and reverse land degradation, and halt biodiversity loss. Vegetable oil constitutes 80% of total human nutrition. Forests, which cover 30% of the world's land surface, help to keep air and water clean. However, earth and life on it are in danger. Arable land is disappearing 30–35 times faster than in the past. Deserts are expanding. Animal breeds are becoming extinct. This Sustainable Development Goal is aimed at preserving and restoring dryland ecosystems by 2030.

Goal 16: Promote peaceful and inclusive societies for sustainable development, provide access to justice for all, and build effective, accountable, and inclusive institutions at all levels. No country develops without peace and justice in it, without human rights or government based on the rule of law. Yet, in some parts of the world, relative peace and justice dominate, and individual countries suffer from armed conflicts, crimes, torture, and exploitation that hinder their development. Peace and justice are the goals for all the world's countries. The aim of this Sustainable Development Goal is to reduce all forms of violence and to suggest governments and communities finding long-term solutions to the conflicts and insecurities. This means strengthening the rule of law, reducing the illegal flow of arms trafficking, and involving developing countries in institutions of global governance.

Goal 17: Strengthen the means of implementation and revitalise the Global Partnership for Sustainable Development. Today, more than ever before, the world is interrelated due to possibly accessible new information technologies, travels, and global institutions. There is a growing consensus that we need to work together in order to halt the consequences of climate change.

767 million people on the planet still live on less than USD 1.90 a day, whereas 793 million people face a shortage of food and drinking water on

a daily basis (United Nations 2017). Our planet needs a more resolute policy to reduce maternal mortality, achieve sustainable energy goals and larger investments in sustainable infrastructure as well as ensure the rights of every child to quality education. It is estimated that if all children, eligible for learning, graduates from high school by 2030, per capita income will rise by 75% till the year 2050, we may win the fight against poverty in a decade (United Nations 2017).

Almost all the Sustainable Development Goals are related to climate change to some extent. This issue has even been marked out by a separate point, namely, Goal 13. Measures to halt climate change are envisaged and, in the cases, when it is impossible to do so, to apply effective adaptation measures.

However, at first glance, in the Sustainable Development Goals, very little attention is paid to the element of culture and the role of culture while forming a sustainable and smart society. The cultural issue is closely related to the issue of climate change, and this possibly new component of sustainable development can be used effectively to achieve the Sustainable Development Goals that are, as already mentioned, either literary or figuratively related to climate change.

Modern society is transforming into a new, sustainable, and environmentally friendly society with the aim of providing humanity opportunities to build safer, healthier, and richer world where all the people have access to constant teaching and learning, new knowledge, and values and the need to know, understand, and make meaningful and responsible actions (Chakori 2017). None of the Sustainable Development Goals is directly culture oriented, i.e. definitions do not use the concept of culture directly. Despite this, it is worth paying attention to several aspects. The fourth Sustainable Development Goal aims to ensure all-encompassing and equal quality education, which can be implemented by acquiring new and relevant knowledge that would help to promote sustainable development. In order to achieve this goal, already existing and possible measurements that encourage citizenship in society and accessibility to cultural diversity by assessing the contribution of culture to sustainable development shall be used.

Another Sustainable Development Goal, which may promote sustainable, inclusive, and sustained economic growth, is oriented towards a comprehensive development policy of the countries with the goal of creating and forming a sustainable economy, which is possible only by supporting productive activities, using innovations as well as personal or group creativity and other related activities.

Goals 9 and 10 refer to the implementation of a concept of sustainable tourism. The sustainable tourism stimulates resilient infrastructure building, promotes inclusive and sustainable industrialisation, and increases opportunities for innovations and their successful implementation not only in large cities as it provides the evaluation of local elements of culture in creating necessary monitoring tools.

2.1.2 Integration of Culture to Achieve the Sustainable Development Goals

Cultural participation has a significant impact on people's quality of life: it contributes to their well-being and promotes a sense of belonging to a society.

The eleventh Sustainable Development Goal, the main objective of which is inclusive, safe, and sustainable cities and settlements, is of particular importance culturally. The achievement of this goal can be realised by the implementation of programmes and strategies that directly oblige countries to protect cultural and natural heritage.

Consequently, cultural aspects play a significant role in achieving the Sustainable Development Goals. Table 2.1 shows links between the Sustainable Development Goals and culture.

The cultural aspect of sustainability creates strong relations and is compatible with the other three dimensions of sustainable development. The conceptualisation of culture as the fourth component of the sustainable development, along with the ecological, social, and economic components, is based on a well-established and simple approach. Such an approach has its own dangers because culture can be viewed in a rather limited way as an artistic and cultural-creative sector (Hawkes 2001). This narrows down the definition of culture. It is important to note that this concept allows culture to be understood both qualitatively and quantitatively. Nevertheless, the role of the fourth component offers many opportunities. In this way, culture can be connected to the overall concept of sustainable development. The introduction of culture as the fourth dimension of sustainable development makes it possible to define the characteristics of sustainable development in the sector of arts and culture. Cultural values can be used in policymaking and sustainable development strategies and are practically applicable in artistic and cultural organisations and businesses. Artistic and creative values can be used, for example, in defining the criteria for assessing the sustainability of a particular policy, organisation, or company and in defining criteria that make it possible to evaluate the contribution of culture to the process of sustainable development and creation of product or image.

Depending on the circumstances and goals, all three cultural roles are important as everything depends on the context. The presented roles are not an evolutionary path to follow, but in this three-role system, certain trends, dynamics, and trajectories can be seen. Policies are becoming more diverse and multilayered, requiring a broader understanding of the sustainable development process, a dialogue, and ongoing interdisciplinary cooperation. It is clear that conceptualising culture as an equivalent component of sustainable development expands the definition of culture, which requires a systemic approach encompassing both natural and human-made worlds.

TABLE 2.1

Links between the Sustainable Development Goals and Culture

SD Goal No	Sustainable Development Goal	The role of culture in implementing the Sustainable Development Goal
Goal 1	End poverty in all its forms everywhere. Indicators: 1. The access to use chosen infrastructures of the cultural community (museums, libraries, media resource centres, exhibition venues for performing arts) compared to the distributions of the country population in administrative units below the state level. 2. Men and women with the ability to reach the main cultural services and resources (libraries, community centres, art centres, museums, local heritage conservation centres, etc.) in 30 minutes by foot.	Cultural services are services that shall be accessible to all, ensuring equal access to them by paying particular attention to poor and vulnerable individuals. The creation of cultural expression, services, goods, and heritage sites may contribute to inclusive and sustainable economic development.
Goal 2	End hunger, achieve food security and improved nutrition, and promote sustainable agriculture.	Traditional knowledge that allows preserving the existing genetic resources should be recognised and maintained in each area. In this way, it is possible to reduce the threat of famine and improve local population nutrition opportunities by using genetic seed diversity and promoting fair payments in agriculture. The sustainable development of agriculture can be promoted by using traditions of cultivation, livestock farming, and horticulture.
Goal 3	Ensure healthy lives and promote well-being for all at all ages.	New health policies and healthy lifestyle promoting programmes must be culturally relevant. Local customs, which can be integrated into traditional healthcare systems, play an important role here. Individual participation in the cultural life of the country can also improve their health and overall well-being.

(Continued)

TABLE 2.1 (*Continued*)

Links between the Sustainable Development Goals and Culture

SD Goal No	Sustainable Development Goal	The role of culture in implementing the Sustainable Development Goal
Goal 4	Ensure inclusive and equitable quality education and promote lifelong learning opportunities for all. Indicators: 1. The percentage of teaching hours for art education, taking into account the total amount of teaching hours in the first two years of secondary school (grades 7–8). 2. The percentage of workers in primary and secondary education with special training in artistic and cultural disciplines. 3. The percentage of primary and secondary schools with a library. 4. The percentage of the population who have participated in cultural activities at least once in the last 12 months. 5. Global Cultural Engagement Index (and related indicators).	Aspects of cultural diversity, art education, languages, and culture that play an important role in sustainable development should be integrated into education programmes of all levels. Cultural approach, including the recognition of local languages and local level skills, and participation of interested cultural parties must be prevailing in creating education programmes at all levels: it complies with human rights and can contribute to the educational objectives, including student motivation and community relations. This goal can be achieved by ensuring that all learners acquire the knowledge and skills needed to promote sustainable development. It requires the use of assessment of the sustainable development and sustainable lifestyle, human rights, gender equality, peace and non-governmental cultural promotion priorities as well as the assessment potential for global citizenship development, cultural diversity, and contribution of culture to the sustainable development.
Goal 5	Achieve gender equality and empower all women and girls.	Gender equality should be achieved in cultural life as well. It requires widening the opportunities for women and girls to take an active part in cultural life and manage their projects and organisations in this field. Cultural practice, which is mainly run by women and girls, must be more visible and acknowledged. Narratives that talk about gender discrimination or show the importance of the role of women and girls in cultural life are necessary.

(*Continued*)

TABLE 2.1 *(Continued)*

Links between the Sustainable Development Goals and Culture

SD Goal No	Sustainable Development Goal	The role of culture in implementing the Sustainable Development Goal
Goal 6	Ensure availability and sustainable management of water and sanitation for all.	Traditional knowledge, values, and customs can provide an opportunity to learn how to use ecosystems properly, which is directly related to clean water availability and sanitation.
Goal 7	Ensure access to affordable, reliable, sustainable, and modern energy for all.	Certain cultural factors can be used as models for energy generation and consumption. By employing traditional businesses and cultural heritage knowledge, creative people can participate in creating educational and awareness-raising activities related to energy generation and consumption, especially in the field of energy efficiency.
Goal 8	Promote sustained, inclusive, and sustainable economic growth, full and productive employment, and decent work for all. Indicators: 1. The percentage of people employed in the cultural sector of the total working population. 2. The percentage of UN development assistance systems, national development plans, and local development plans integrating culture. 3. Contribution of creative and cultural activities to the gross domestic product. 4. Index of interface and coverage of technical and vocational training and higher education systems in the field of art and culture. 5. The percentage of countries that implemented/adopted social security and tax laws and regulations for self-employed artists according to 1980 UNESCO recommendation on the status of an artist.	Cultural and creative sectors can become areas of inclusive, sustainable, and fair employment if decent working conditions, which are in line with international human rights, are ensured. Cultural aspects can be integrated into tourism strategies. It is essential to ensure that cultural identities, as well as related activities and assets, would not be deco-desexualised and that the benefits would be reinvested into cultural activities. This goal can be achieved by supporting productive activities, job creation, entrepreneurship, creativity, and innovations in the development of small- and medium-sized enterprises, taking advantage of local access, human resources in the regions and isolated areas, and adequate funding. Sustainable tourism promotes local culture, which, in turn, creates jobs and culturally promotes the development of valuable products.

(Continued)

TABLE 2.1 ***(Continued)***

Links between the Sustainable Development Goals and Culture

SD Goal No	Sustainable Development Goal	The role of culture in implementing the Sustainable Development Goal
Goal 9	Build resilient infrastructure, promote inclusive and sustainable industrialisation, and foster innovation. Indicators: 1. The number of countries that implement the national strategy of creative industry development.	If the cultural infrastructure is properly designed, it ensures affordable and equal access to cultural life as a part of the high-quality, reliable, sustainable, and resilient infrastructure that should be accessible to all. Artists and creative individuals can participate in processes of research, development, and innovations in various industrial areas.
Goal 10	Reduce inequality within and among countries.	The participation of individuals in cultural life contributes to reducing inequality since the participation must not be restricted by age, gender, disability, race, ethnicity, religion, economic, or another status. Artists and creative individuals have the opportunity to participate in the process of creating and presenting narratives that give an exclusive status to developing countries. All approaches to migration should include the cultural aspect and intercultural dialogue.

(Continued)

TABLE 2.1 ***(Continued)***

Links between the Sustainable Development Goals and Culture

SD Goal No	Sustainable Development Goal	The role of culture in implementing the Sustainable Development Goal
Goal 11	Make cities and human settlements inclusive, safe, resilient, and sustainable. Indicators: 1. The percentage of national and local urban development plans that include a specific "cultural impact assessment". 2. The number of identified cultural and natural heritage objects (objects and artefacts). 3. The number of dangerous natural and cultural heritage objects. 4. The number of public libraries per 1,000 inhabitants. 5. The percentage of the budget allocated to the preservation of cultural and natural resources. 6. Multidimensional system development index of heritage sustainability. 7. The part of cities with the integrated urban policy that protects cultural and natural heritage. 8. The part of urban land for open public spaces (streets, squares, gardens, parks, etc.). 9. The part of urban land for public protected premises (libraries, museums, etc.).	Many important objects and elements of tangible and intangible cultural heritage are located in cities and play an important role in sustainable local development; indeed, cultural aspects are essential in promoting sustainable local development. Green and public spaces can contribute to the development of local cultural activities and must be accessible to all. Traditional construction methods, knowledge, and local materials can provide information on the renovation of existing buildings and the design of new buildings. Cultural factors inform about behaviour in cities, including transport and mobility, usage of environment, etc. The implementation of this goal enables the strengthening of efforts in the area of global cultural and natural heritage protection by ensuring universal access to safe, inclusive, and accessible green and public spaces, particularly for women and children, the elderly and the disabled.
Goal 12	Ensure sustainable consumption and production patterns. Indicator: 1. The percentage of national and local sustainable tourism development strategies involving the culture section.	Local and traditional products that are suitable for promoting sustainable consumption must be recognised and valued in order to achieve the goal.

(Continued)

TABLE 2.1 (*Continued*)

Links between the Sustainable Development Goals and Culture

SD Goal No	Sustainable Development Goal	The role of culture in implementing the Sustainable Development Goal
Goal 13	Take urgent action to combat climate change and its impacts. Indicator: 1. The percentage of national and local climate change strategies that consider the role of the cultural aspect of promoting environmental sustainability.	Intercultural activities and traditional knowledge of environment-friendly practices must be explored and promoted. Cultural professionals have the opportunity to inform the public about climate change and its consequences through cultural activities.
Goal 14	Conserve and sustainably use the oceans, seas, and marine resources for sustainable development.	The goal can be implemented by identifying and strengthening certain cultures that maintain traditions related to the preservation of marine and coastal ecosystems.
Goal 15	Protect, restore, and promote sustainable use of terrestrial ecosystems, sustainably manage forests, combat desertification, and halt and reverse land degradation as well as halt biodiversity loss.	Cultural factors related to the preservation of terrestrial ecosystems including relevant local and traditional knowledge must be included in the preparation, implementation, and evaluation of policies and programmes in this field.
Goal 16	Promote peaceful and inclusive societies for sustainable development, provide access to justice for all, and build effective, accountable and inclusive institutions at all levels. Indicators: 1. Laws ensuring the right to receive information from public institutions, based on international standards. 2. Legal regimes ensuring compliance with international standards on freedom of expression, association, and assembly. 3. The percentage of libraries regularly providing specific training on media and information literacy competences in order to help users access and use information.	In order to implement this Sustainable Development Goal, nationalised and otherwise expropriated cultural objects should be returned to the concerned communities. Citizens should be able to participate in the preparation, implementation, and assessment of cultural policies and programmes. Cultural objects, including libraries and media centres, promote access to the information. Strategies that condemn violence and promote peace shall also include a cultural component.

(*Continued*)

TABLE 2.1 (*Continued*)

Links between the Sustainable Development Goals and Culture

SD Goal No	Sustainable Development Goal	The role of culture in implementing the Sustainable Development Goal
Goal 17	Strengthen the means of implementation and revitalise the Global Partnership for Sustainable Development.	The cultural aspect should be integrated into international, national, and local sustainable development strategies seeking to implement the 2030 Agenda. The capacity of cultural stakeholders to address the challenges of sustainable development should be strengthened, whereas capacity building should also enable other sustainable development groups to understand the importance of cultural aspects. The capacity of cultural stakeholders to produce and distribute cultural goods and services, particularly those representing lesser-known cultural expressions, should be increased.

Source: Created by authors.

2.2 The Link between Sustainable Development and Climate Change

A close link exists between sustainable development and climate change. Since their main connection is energy, it is necessary to examine how policies of sustainable development and climate change can be identified and implemented together. A system of indicators is necessary to establish a link among the Millennium Development Goals (which are the key targets of sustainable development policy), national development, and climate change mitigation programmes.

In the Third Assessment Report by the Intergovernmental Panel on Climate Change (IPCC), a chart linking climate change, natural systems, people, and opportunities for socio-economic development was introduced.

In Figure 2.1, the interaction between the climate change mitigation policy and adaptation efforts is presented.

Figure 2.1 vividly illustrates how paths of socio-economic development affect GHG emissions as well as climate change, which leads to changes in human and natural systems.

Following the IPCC report, several studies had emerged that addressed climate change and sustainable development in a comprehensive manner. These studies aimed to establish a clear divide between natural processes

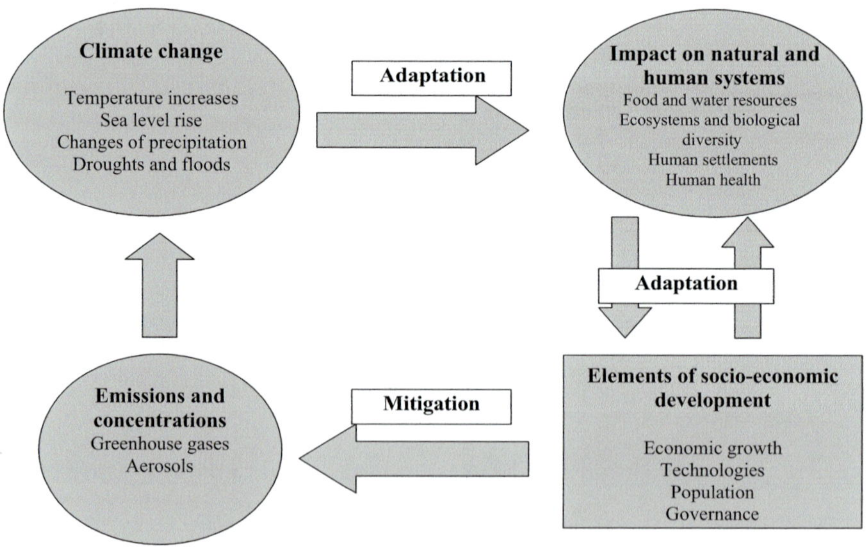

FIGURE 2.1
Interaction among policies of sustainable development, adaptation, and climate change mitigation.

Source: Created by authors.

with their outcomes and their relation to human and social activity outcomes, which are determined by the implementation of policy provisions. These policies include direct climate change mitigation measures as well as more general policies that impact vulnerability due to climate change together with adaptation and mitigation capabilities.

Furthermore, we will examine the options for sustainable development, climate change, adaptation, and mitigation as well as their links in more detail. When it comes to climate change mitigation and adaptation, it must be recognised that these processes are influenced by the trajectory of development together with institutions implementing it and by the specific options available for climate change mitigation and adaptation. This means that policy-driven development trajectory and institutions implementing it, despite their broader development targets, have an indirect impact on climate change adaptation and mitigation.

This impact can be both positive and negative. Several studies had been developed that proposed integrating climate change mitigation and adaptation into development policies to ensure more sustainable development (Munasinghe 1991).

Kenneth Joseph Arrow proposed a method to assess alternative options for consumption growth in the long run from a sustainable development perspective. Intermediate consumption and usefulness are used as criteria for assessing sustainable development. One of the determinants of consumption and usefulness is the productive base of society, which consists of capital (productive, human, natural) and institutions. Institutions can be understood as a part of the capital as well as a guide for resource (and capital) distribution. Institutions cover the legislative framework, formal and informal markets, various state agencies, interpersonal networks, and the rules of conduct and the norms on which they base their activities (Arrow et al. 2012).

From the perspective of sustainable development, it is important how and to what extent policies can affect the productive base of society, while ensuring that the development trajectory is more sustainable when implementing certain adaptation and mitigation policies.

Policy on climate change mitigation or adaptation can be considered as a starting point, while sustainable development can be seen as a result of the indirect influence of this policy. This approach is focused on more specific sectoral policies or climate change policy instruments that ensure adaptation and mitigation targets but are not necessary for all dimensions (social, economic, and environmental) of sustainable development. This shows that climate change policy will be held back by the compatibility of development goals with the global environment. Besides, policies that fail to assess the economic and social consequences may remain unsustainable for a long time.

The next figure presents a chart of interaction among climate, natural systems, and society as well as the impact of policies among themselves.

Figure 2.2 shows three systems represented by ellipses: climate change, natural system, and socio-economic system. GHG emissions affect the climate

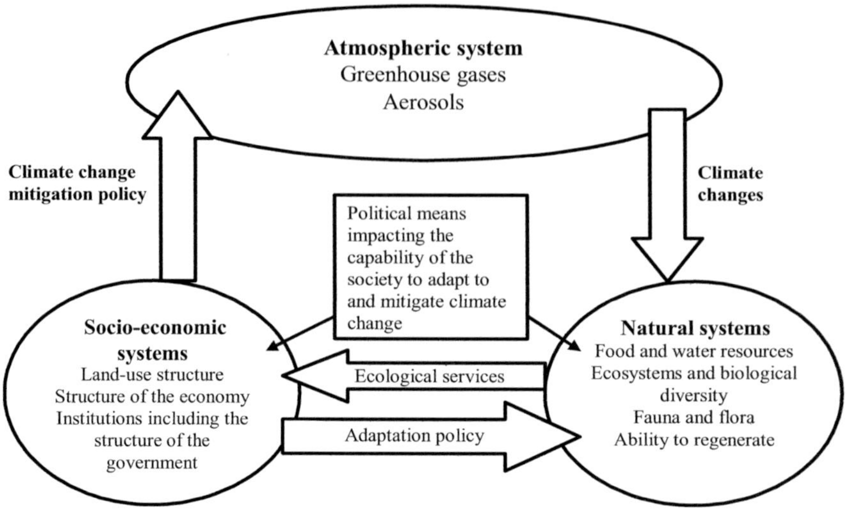

FIGURE 2.2
Interaction among climate, natural, and socio-economic systems and their connection with climate change adaptation and mitigation.

Source: Created by authors.

system and cause climate change, which impacts the natural system again. The impact of climate change on the natural system depends on the vulnerability of the system and its ability to resist those impacts. As a result, ecological services are provided to the public such as agricultural land, fishery resources, water resources, etc. It should be stressed that the mutual influence of all systems is complex and non-linear, making it difficult to assess all aspects of the links among them.

The response to climate change depends on the productive base of society, which includes capital, institutions, and adaptation and mitigation capabilities. Policies aimed at ensuring the productive base of society and the capacity for adaptation and mitigation are presented in Figure 2.2 and grouped in one division. In addition, the policies that are divided here have many synergistic effects, which together affect the natural system and the achievement of Sustainable Development Goals.

In this conceptual framework, adaptation and mitigation policies are treated as more isolated policy efforts such as investments in protective measures (dams, new types of crops, technologies to reduce GHG emissions, carbon taxes, etc.), which only affect the marginal response of the production base of society. The factual outcome of specific mitigation and adaptation policies depends on adaptation and mitigation capacities that are part of the social, economic, and natural state of systems. Together, the results of adaptation and mitigation policies depend on the propulsion of social and economic development such as investments, technologies, population, governance, and environmental priorities.

> Dual-link: climate change → sustainable development and sustainable development → climate change

There is a dual-link between climate change and sustainable development. On the one hand, climate change affects the main natural and living conditions together with the social and economic development foundations, but, on the other hand, society's sustainable development priorities impact both vulnerability and GHG emissions.

Many discussions on the links between sustainable development and climate change were initiated in the IPCC Third Assessment Report. The synthesis report stressed the importance of understanding the links between climate change and sustainable development and concluded that climate change issues are a part of the broader challenges of sustainable development. As a consequence, climate change policy can be much more effective if it is consistently integrated into the wider regional and national sustainable development strategies since the climatic diversity and changes, the response to climate change policies and the related socio-economic development will affect the country's capacity to implement Sustainable Development Goals. And vice versa, the realisation of the Sustainable Development Goals will impact the potential of climate change policies and the success of their implementation. The socio-economic and technological characteristics of different development trajectories will have a huge impact on GHG emissions, climate change and its influence, and the capacity to adapt and mitigate climate change.

The key findings of the IPCC are more conceptual and confirm that sustainable development can serve as a framework for understanding society's ability to respond to climate change.

Climate change will also affect sustainable development: due to the negative impact of climate change on ecosystem services, human health, agricultural production, and many other areas, it will be more difficult to implement Sustainable Development Goals. The impact of climate change on development perspectives has been described as a comprehensive poverty reduction and climate change mitigation project "Climate change will complement existing poverty". The negative effects of climate change will be stronger in developing countries, as the latter have limited capacity to adapt to climate change. In these countries, the poorest sections of society with the least resources and the least capabilities to adapt will be the most vulnerable. Predicted changes of extreme climate events in their frequency, power, and intensity as well as average climate changes will mainly affect their quality of life and continue to increase inequalities between the developing and developed world. Thus, it can be stated that climate change is a strong source of poverty growth.

By recognising the dual-link between sustainable development and climate change, it must be stressed that policies shall simultaneously cover both sustainable development and climate change mitigation and adaptation targets.

2.2.1 Development Targets and Sustainable Development Tasks

The concept of sustainable development had been long discussed in the theoretical literature until it became the most significant target in economic, social, and environmental development of Agenda 21, UN institutions, national governments, and private individuals. There are many definitions of sustainable development, but here we will use a significantly pragmatic perception of sustainable development as the goal of human well-being both from a short- and long-term perspective.

The simplest concept of sustainable development is defined as co-development ensuring economic, social, and environmental development targets. These targets can be expressed by a number of economic, environmental, human, and social indicators, while the impact of the implemented policies on sustainable development can be assessed both quantitatively and qualitatively. Table 2.2 provides examples of the main dimensions of sustainable development.

As a starting point for the links between sustainable development and climate change, basic Sustainable Development Goals such as health, education and energy, food, and access to water can be achieved through good governance and without impacts of climate change. Thus, the Sustainable Development Goals are addressed together with the current development targets and the challenges of achieving them.

Using such a pragmatic approach to the concept of sustainable development requires the acknowledgement that the main conceptual assumptions are based on the key development paradigms and analytical approaches which are used in many sustainable development studies. The perception of development goals and the discovery of a compromise among different policy targets depend on the applied development paradigm. Therefore,

TABLE 2.2

Examples of Economic, Environmental, Human, and Social Dimensions of Sustainable Development

Economic Dimensions	**Human Dimensions**
Economic growth	Training and education
Investments	Health
Technological changes	Gender
Revenue in specific areas	
Environmental dimensions	**Social dimensions**
Atmospheric pollution	Governance
Water pollution	Income distribution
Waste	Participation
Biodiversity	Justice
Depleting resources	

Source: Created by authors.

in this chapter, we will show how policy recommendations on sustainable development and climate change depend on alternative approaches to the development itself.

A system of sustainable development indicators aimed to measure policy decisions at the lowest and highest levels, called Action Impact Matrix (AIM), was developed by Mohan Munasinhge. AIM links national development targets and programmes with the assessment of sustainable development indicators and the involvement process of governmental authorities, academic and civil society as well as the private sector.

Alternative development paradigms: The development targets of different paradigms, as well as different scientific disciplines, are treated differently. The development approach in paradigms is the basis on which the key elements, necessary to be evaluated in order to establish the link between development and climate change and its dynamics, are determined. Further, we will examine many development paradigms and their impact on climate change mitigation studies.

Paradigms based on economic theories usually define a number of goals that are important contributions to the well-being of people. In this case, development goals can be perceived as functional elements of a certain value. The value function reflects the results of individual utility by using various goods and services.

Some economic paradigms rely on the economic welfare function, where efficient resource allocation is approached as in neoclassical economics, while deviations from this state and ways of overcoming these deviations are not addressed. From the point of view of these paradigms, the link between sustainable development and climate change is approached as a simple statement that climate change mitigation only leads to additional costs for optimal status.

Meanwhile, other economic-based paradigms, such as institutional economics, are more focused on formulating a question: how do markets or other information-sharing mechanisms provide the basis for economic interaction? In this context, the main idea is how mitigation policies can be integrated into the institutional framework of the economy. Thus, the institutional approach to climate change analysis involves the assessment of the work of existing institutions and making proposals on improving their work by assessing market weaknesses and limited capacity. Furthermore, institutions include governance and political systems that are the main factors for determining development trajectories.

For example, Partha Dasgupta recommends studying the allocation of resources as a counterpoint to the effects that can be measured by parameters of well-being. Possibility to get an income and meet basic needs (education, food, energy, medical services) is considered the basis of the human being (Dasgupta 2000, 2003).

In the context of climate change mitigation policies, it should be examined whether and to what extent, policies may impede or facilitate access of individuals to specific resources and freedoms. In some cases, mitigation policies may prove to be too expensive if more expensive types of energy carriers are introduced, while, in other cases, they may increase the access to energy if energy conservation measures are introduced and energy becomes more affordable.

This approach is close to the paradigms of social sciences, which are based on the ethics of liberalism or equalisation and emphasise the rights of individuals to participate in decision-making and goal achievement processes. These theories shape the assurance of individual freedoms, so issues such as the link of risk and justice with climate change will be addressed by studying the local participatory process.

Development trajectories: When analysing possible development trajectories, it is very important to recognise that several different definitions of sustainable development exist based on the concept of sustainable development established by the Brundtland Commission.

If we regard sustainable development as the policy target, then the important question is how policies can ensure a more sustainable development trajectory. Society's development trajectory is exposed to many important decisions relating to investments, use of natural resources, lifestyle and consumption, selection of technologies as well as the institutional structure, which leads to the base conditions of these choices. All of this can be achieved through sustainable development policies.

Sustainable development policies cannot be strictly separated from other policies; however, a variety of policy recommendations, which cover different components, can still be discussed. They include many policies related to nature conservation, laws regulating access to resources, environmental taxes, promotion of organic agriculture, the increasement of human and institutional capital, research and development, financial schemes, and technology transfers. These policies are usually not implemented as a part of the general sustainable development policies package, but as policies, dedicated to addressing specific policy targets such as air pollution standards, organic food and health issues, GHG emission reduction, income generation for specific groups of people, or the development of green technology industry. In this way, the development trajectory is obtained as a result of many economic and social transactions initiated by governmental policies, private sector initiatives, and choices of consumers.

There are a great number of scientific studies, which are based on economic theories and complex systemic methodology, ecological science, as well as other methodologies and determine the conditions under which the development of the trajectory meets the sustainable development requirements. Arrow summarised the controversy of discussions between economists and ecologists concerning the nature of current development and its compliance with the criteria of sustainable development. Economists are looking into the ability of the economy to maintain an adequate standard of living and state that welfare should be optimised to ensure that current consumption is not excessive. However, it is difficult to determine the optimal level of current consumption; thus, the theoretical debate revolves solely around the factors leading to development that is not sustainable. Those factors include the link between market return on investment and social discount rates and the link between market prices for goods and their social costs.

2.3 Implementation of Sustainable Development and Climate Change Policies

The impact of climate change policy on development policies: Studies aimed at evaluating the impact of sustainable development on climate change and vice versa explore a number of current development challenges. Here, we will overview international policy initiatives and solutions that include development goals as well as discuss how climate change issues can be addressed in conjunction with development goals. We will also look at many examples of how countries that have formulated development goals also consider mitigation aspects of sustainable development.

Many international initiatives, including the World Summit on Sustainable Development 2002 (WSSD) and the actions of the Commission on Sustainable Development, have set many different goals for different sectors related to climate change mitigation policies. We will take a look at these initiatives below.

The main decision of the World Summit on Sustainable Development was the confirmation of the Water, Energy, Health, Agriculture, and Biodiversity (WEHAB) initiative's connection with sustainable development (World Summit on Sustainable Development 2002). The WEHAB sectors reflect the main areas in which WSSD countries consider it necessary to establish policies in order to implement Agenda 21. The justification document prepared for the WEHAB presents numerous policy measures in different sectors that relate to climate change issues. Below a description of the main sectors and policies is presented.

Water: Ensuring safe water supply, water management, agricultural efficiency, human health, disaster preparedness, financial resources, institutional and technical capacity, and protection of water systems.

Energy: Accessibility, efficiency, renewable energy, advanced organic fuels, and transportation.

Health: Reduction of poverty and malnutrition, access to health services, reduced infant, child and maternal mortality, control and eradication of major diseases, planning, link with the environment, risk management capacity, and crisis preparedness.

Agriculture: Increasing productivity and sustaining the natural foundation, knowledge generation and information transferring, partnerships between public and private sectors, and political and institutional reforms.

Biodiversity: Integrating sustainable development into economic and sectoral development plans, and restoring and protecting biodiversity.

Aspects of climate change policy can be linked to the Millennium Development Goals (MDGs). Commission on Sustainable Development stressed that climate change could only worsen the situation of the poor and hinder the implementation of the Millennium Development Goals. SDS has proposed to complement the MDGs with energy development tasks to strengthen the connection between the MDGs and climate change mitigation.

MDGs could be linked to energy, food, water access, and climate change impacts as well as vulnerability and adaptation (Table 2.3).

The basis for the joint implementation of sustainable development and climate change policies is illustrated in Figure 2.3. It consists of the general state of the natural environment, the socio-economic system, and policy priorities. The latest international trends in sustainable development and management of climate change are the increasing involvement in these processes by various institutions and other stakeholders. These stakeholders include international agencies, global forums, private companies, and non-governmental organisations. The involvement of various stakeholders in this process creates new opportunities to combine international agreements and commitments with voluntary action and market-driven processes. Several new initiatives by private efforts are to combine green business, sustainable development, and climate change mitigation.

Many companies have included voluntary commitments in their development strategy reflecting their social responsibilities and environmental goals that go beyond their traditional responsibilities. The majority of companies voluntarily commit to reducing GHG emissions. Several international networks are promoting these initiatives: World Business Council for Sustainable Development (WBCSD), The Climate Group, etc. These initiatives serve as platforms for collaboration among companies, non-governmental organisations, governments, and other stakeholders.

WBCSD has identified the following main targets for private companies in order to mitigate energy and climate change:

- to ensure long-term success in a competitive business environment;
- to achieve the objectives of national and regional climate change mitigation policies;
- to assess risks and economically effective options for reducing pollution; and
- to participate in the innovation process of processes and products.

TABLE 2.3

The Link among the Millennium Development Goals, Energy, Food, Access to Water, and Climate Change

Goals	Sectoral Subjects	Relations to Climate Change
From 1990 to 2015, to reduce the proportion of people whose income is less than 1 USD a day	Energy: • Energy supply to local businesses • Facilitation of income generation • Energy supply to the equipment • Employment related to energy supply Food/water: • Increment of food production volumes • Improved water supply • Employment	Energy: • GHG emissions • Increased adaptive capacity due to higher incomes and reduced dependence on natural energy sources Food/water • GHG emissions • Increased agricultural productivity can reduce climate change • Better water supply management can help adaptability
From 1990 to 2015, to reduce the proportion of people who suffer from hunger	Energy: • Energy supply for machinery and irrigation equipment in agriculture Food/water: • A more efficient production process which increases production volumes and reduces waste • Land and food distribution	Energy: • GHG emissions Food/water: • Increased GHG emissions from some agricultural activities; however, it is partly offset by better use of agricultural waste • The ability of farmers to adapt depends on the income and the form of land ownership

(*Continued*)

TABLE 2.3 (*Continued*)

The Link among the Millennium Development Goals, Energy, Food, Access to Water, and Climate Change

Goals	Sectoral Subjects	Relations to Climate Change
To ensure that by 2015 children will have access to primary education everywhere	Energy: • Reduced time to provide energy • Electricity for reading • Electricity for educational information means, including TV and computers Food/water: • Shorter time in this sector allows devoting more time for education • Better health increases children's ability to read	Energy: • Education can strengthen the ability to mitigate climate change Food/water: • Education can strengthen the ability to mitigate climate change
To ensure that girls and boys have equal access to primary and secondary education by 2005 and to the education of any level by 2015 at the latest.	Energy: • Due to modern energy services young women or girls no longer need to provide themselves with energy and they have more free time • New electronic educational tools are making it easier for girls to access information at home Food/water: • Modern agricultural production technologies and improved water supply	Energy: • Education can strengthen the ability to mitigate climate change Food/water: Education can strengthen the ability to mitigate climate change

(*Continued*)

TABLE 2.3 (*Continued*)

The Link among the Millennium Development Goals, Energy, Food, Access to Water, and Climate Change

Goals	Sectoral Subjects	Relations to Climate Change
From 1990 to 2015, to reduce the mortality rate of children under five by two-thirds	Energy: • Energy supply can speed up the development of healthcare facilities • Reduced air pollution from sources that use traditional fuel for combustion • Reduced time spent on collecting fuel may increase time spent on children's healthcare Food/water: • Improved health due to increased supply of high-quality food and clean water • Reduced time spent on providing food and water can increase the time spent on children's healthcare	Energy: • GHG emissions Food/water: • Improved health increases the ability of adaptation
From 1990 to 2015, to reduce maternal mortality by three quarters	Energy: • Electricity supply to health clinics • Reduced air pollution from sources that use traditional fuel for combustion and other health improvements Food/water: • Improved health due to increased supply of high-quality food and clean water • Reduced time spent on providing food and water can increase the time spent on children's healthcare	Energy: • GHG emissions Food/water: • Health improvement increases the vulnerability of climate change and the ability of adaptation

(*Continued*)

TABLE 2.3 (*Continued*)

The Link among the Millennium Development Goals, Energy, Food, Access to Water, and Climate Change

Goals	Sectoral Subjects	Relations to Climate Change
AIDS, malaria, and other serious diseases	Energy: • Electricity for health facilities • Cooling of vaccines and medical devices Food/water: • Improved health due to cleaner water supply • Manufacturing technologies of the food industry that reduce the likelihood of malaria	Energy: • GHG emissions from the increased health clinic service; however, health improvements may reduce the need for health clinics Food/water: • Health improvement increases the vulnerability of climate change and the ability of adaptation
To stop the unsupported exploitation of natural resources	Energy: • Deforestation allows waste fuel to be collected • The usage of non-renewable resources Food/water: • The decline in land ownership	Energy: • GHG emissions • Separation of carbon Food/water: • Separation of carbon • Better conditions in agricultural production increase adaptability
From 1990 to 2015, to reduce the proportion of people who cannot access drinking water	Energy: • Electricity for pumping and distribution systems Water: • Better water systems	Energy: • GHG emissions Water: • Increased vulnerability and adaptability
To expand global cooperation in development		

Source: Created by authors.

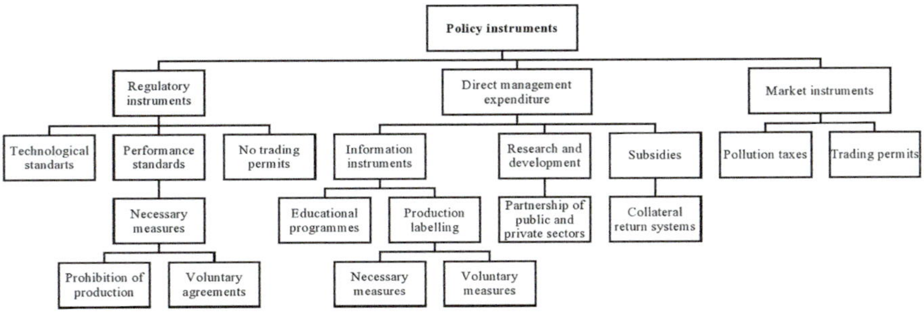

FIGURE 2.3
Measures of the climate change policy.

Source: Created by authors.

The WBCSD has prepared guidelines for monitoring and accounting for GHG emissions under IPCC GHG inventory standards. The WBCSD also emphasised the importance of accounting for GHG emissions in order to demonstrate eco-efficiency and to maintain the company's image of accountability and transparency. In addition, the implementation of this GHG accounting methodology allows companies to participate in emissions trading.

Moreover, insurance companies have their own rules of operation which include activities for ensuring sustainable development and climate change (Munich RE initiative). An example of a non-governmental organisation's initiative combining the goals of sustainable development and climate change mitigation is The Gold Standard prepared by the WWF. The Gold Standard is designed to assess how CDM projects can support the broader Sustainable Development Goals and if they are not included in their GHG reduction tasks.

Climate change mitigation measures: Each country chooses individual climate change mitigation measures. On the basis of the IPCC Third Assessment Report, the following chart of all the climate change mitigation measures is drawn up. Thus, the main climate change mitigation policies can be grouped as follows:

- market-based mechanisms (pollution taxes, emission allowances, subsidies, and collateral return systems);
- regulatory instruments (emission permits, operating and product standards, and prohibitions);
- voluntary agreements; and
- direct government expenditure on research and development (or subsidies/financing of information tools, including educational programs and eco-labelling schemes).

Each of these climate change mitigation measures has its own disadvantages and advantages when comparing with others. The IPCC Third Assessment

Report indicates that various policies that are not related to climate change could influence the dynamics of GHG emissions. These may include structural reforms, for example, the pollution control in the energy sector that is trying to reduce the negative impacts on human health and afforestation for economic or biodiversity's preservation reasons or for salinity control reasons.

The main feature of market-based instruments is that they apply economic incentives to pollution control through pollution taxes and emissions trading allowances. They encourage polluters to adjust to market prices or pollution levels.

Regulatory instruments form the technological and performance standards and project the fuel that is used or the fuel that cannot be used. These types of instruments are applied at a national, sectoral, or company level. They are typically introduced before the market-based instruments and create opportunities for institutional capabilities in the spheres of policy evaluation, monitoring, and enforced implementation.

Voluntary agreements are agreements between a governmental body and a private company, or a unilateral agreement recognised by a public authority. They differ from the regulatory instruments because they are not mandatory and there are no penalties for non-compliance. They demonstrate a mutual responsibility of the contracting parties and are applicable internationally and nationally; they also have low transaction costs and flexibility in implementation.

Information instruments help to overcome the market disadvantage related to the asymmetry of information. Two main information instruments can be used both nationally and internationally: educational programs and eco-labelling. Educational programs are designed to fill knowledge gaps that hinder climate change risk assessment. They also introduce pollution reduction actions such as energy-saving initiatives.

Environmental labelling programs, whether mandatory or voluntary, send signals to users about the characteristics of the equipment (energy efficiency of the equipment). An example of voluntary labelling is the labelling of organic products or ethical trademarks.

The third group of information instruments has recently been identified – informing the industry of long-term indicative goals of the government. For example, the long-term goals for reducing GHG emissions or the tasks of renewable energy sources in the primary energy structure.

An alternative way to tax pollution is the subsidisation of less polluting fuels or energy efficiency improvement measures. Both methods have an impact on energy prices and are favourable for technologies that emit less GHG emissions as well as for clean fuels. Subsidies and pollution taxes are sometimes combined when the taxes for the sold energy are used to subsidise renewable energy.

The following criteria shall be met to select appropriate climate change mitigation measures:

- environmental effectiveness looking from a long-term perspective;
- environmental effectiveness in terms of transaction, information, and forced implementation costs;
- impact on resource redistribution;
- the possibility of administrative and political implementation;
- the dynamic effect related to the learning, innovation, and technological development processes;
- the other economic effect (income redistribution); and
- the other environmental impact (the assurance of ambient air quality).

Countries can select several policies or use them in succession, for example, to introduce technological standards first and then – performance standards. Finally, they could introduce market-based instruments.

Promotion of the spread of innovations and new technologies is a very important means of climate change mitigation. Usually, the government uses subsidies, research funding programs, green purchasing of environmentally friendly technologies as well as a complex of climate change mitigation measures. That way the strengths of the various instruments can be combined, and their negative impact reduced.

Climate change mitigation policy and associated costs include a range of measures used for market development and broader institutional policies (Table 2.4).

When defining climate change mitigation policies, certain assumptions are made in the energy sector (Table 2.5). These assumptions are very important when choosing climate change mitigation policies and justifying the expediency of using the proposed instruments. The most important assumptions are based on population, economy, energy demand growth forecasts, the elasticity of prices and services, discount rate, etc.

Table 2.6 shows the relation between climate change policy and the different dimensions of sustainable development. As can be seen, climate change also has a direct impact on economic growth opportunities and many important social factors as well as political and social issues. In addition, the effects of climate change can be identified domestically and globally.

The most significant thing when selecting the climate change policy and measures for its implementation is to take into account its impact on sustainable development and to harmonise climate change policies with those of sustainable development; otherwise, the climate change mitigation can have even more negative effects than climate change itself. This is particularly the case with developing countries, which are the most vulnerable to climate change but have limited possibilities to adapt to it. Therefore, in these countries, climate change mitigation policies need to be integrated into their country's sustainable development strategies, and their implementation should not hinder the achievement of the MDGs.

TABLE 2.4

Examples of Policy Measure Implementation Related to Climate Change

Market Measures	An Example of a Policy Measure
The development of the market during the transitional period, possibly involving the public sector	Temporary support for special demonstration projects
Privatisation with precise identification of property rights and individual requirements	Land ownership rights
Competition regulation by introducing more market players	Information companies, flexible loans for the expansion of technologies that use renewable resources
Environmental taxes	Carbon taxes
Support for efficiency through savings and stimulation of investments	Support by the financing mechanisms
Publishing of technical standards within the relevant period	Standards for the effective use of electrical appliances
Price liberalisation, support for the international competition	Change in the rate of devaluation, elimination of the subsidies
Redirecting Policy towards Flexibility and Constraint of Acknowledged Technical Systems	
Time selection of infrastructure investments	Long-term planning of energy production and transmission
Subsidies for capital turnover projects	Special capital grants
Subsidised credits for research development	Demonstrations and research programmes
Coordination of specific climate change mitigation measures and their integration into the common investment policy	Information, subsidies
Established (Institutional) Policy Measures	
Establishment of monitoring and the enforced implementation systems	Reporting systems
Established and imposed property rights	Land reforms
A structure to mitigate the risk and/or to reduce the accumulated risk (especially in the capital market) is established	A balanced market
Establishment of specialised organisations to reduce uncertainty and transfer information	Insurance schemes
To implement an international mechanism for technology "transfers"	Clean Development Mechanism
Policy Measures Concerning Human Opportunities	
Learning and educational activities	Opportunity development programmes
The improvement of decision-making processes	Participation, local government
Educational programmes	Providing energy and transport for schools

Source: Created by authors.

TABLE 2.5

Assumptions for the Climate Change Mitigation Policy in the Energy Sector

Assumptions Are Introduced	Meaning and Relevance
Population	Growing population increases GHG emissions
Economic growth	A growing economy increases energy consumption and makes it possible to increase investments, which accelerate the growth of the energy-using equipment
Energy needs • Structural change • Technological change • Lifestyle	Different sectors have different levels of energy consumption intensity; structural changes have a major impact on overall energy consumption Different assumptions in the areas of the intensity of GHG emissions and the use of resources can be used for alternative scenarios
Energy supply • Presence of technology and prices • New technologies • Learning	Possibility for fuel and technology replacement Electricity prices that make alternative electricity supply possible Technology costs related to time, market scale, and institutional competence Opportunities for the introduction of new technologies
Elasticity of prices and revenue	Relative changes in the energy needs due to the changes in prices or revenue; higher elasticity results due to larger changes when energy is used
Implementation and contractual prices	The scope of implementation, regulatory framework, institutional competence, administration
Discount rate	Time The potential price of the capital Assumptions of risk Uncertainty
Policy instruments and regulation • Instruments • Obstacles	Economy against the regulatory measures Implementation costs including the cost of overcoming obstacles, the form of institutional aspects or market trends, including building opportunities and institutional reforms; functioning assumptions
Existing tax system and tax refund	Carbon tax refunds; replacement of distorting taxes
Additional benefit	Integration of local and regional policy measures into many production cases Secondary benefit Targets of social policy: income distribution and employment

Source: Created by authors.

TABLE 2.6

The Impact of Climate Change on Different Dimensions of Sustainable Development

Dimensions of Sustainable Development	Impact of Climate Change	
	Domestically	Outside the Country
Economy	Increased vulnerability of agricultural technologies enhances inequality	As the negative effects of climate change in developing countries increase, inequality increases as well
Health	The poorer people suffer from lower general health standards and reduced access to healthcare	Effects of floods and diseases are more important in developing countries
Social security	All sectors are affected but those that depend mainly on natural resources are affected the most	Greater impact in developing countries
Gender	As the main users of natural resources, women are individually affected by climate change	Economic gender inequality increases
Access to the welfare of society	Government cost-sharing to mitigate the effects of climate change impacts everyone, but it will mostly affect the poor	Adaptation costs are higher in poor countries
Political and social freedoms	With the possibility of social destruction, freedoms can be ruined	The impact of migration can be felt in all countries

Source: Created by authors.

2.4 Sustainability Assessment Methods

Sustainability assessment methods can be divided into four key groups: indicators and indices, sustainability assessment means at the level of products (production methods) as well as at project, and country levels. All these means can be further subdivided based on their sustainable development dimensions (environmental, social, economic, integrated, and including all sustainable development dimensions) (Table 2.7).

2.4.1 Indicators and Indices

The first group for sustainability assessment consists of *indicators*. Indicators are a simple tool that allows assessing the economic, social, and environmental development goals of a country. If environmental, social, and economic indicators are integrated into a single indicator, they make an *index*. Indicators must show the following features: simplicity, wide coverage, and an ability to perform a quantitative assessment that helps to determine tendencies. Tendency assessment allows you to make short-term forecasts.

TABLE 2.7

Sustainability Assessment Methods

	Indicators/Indices	Products, Technology Assessment	Project Assessment	Sectoral, Country Assessment
Environmental dimension	Environmental pressure indicators Ecological footprint	Life-cycle assessment Material input per unit of service Material flow analysis Energy flow analysis Exergy analysis Emergy analysis	Environmental impact assessment Ecological risk analysis	Environmental extended inter-branch balance Inter-branch energy balance Strategic environmental impact assessment Regional emergy analysis Regional exergy analysis
Economic dimension	General national production volume	Life-cycle costs	Accounting of all life-cycle costs	Economic material flow analysis Economic flow analysis Economic inter-branch balance
Social dimension	Social indicators		Social impact assessment	Social inter-branch balance
Integrated method	Human Development Index Environmental Sustainability Index Welfare Index Sustainable national income Genuine progress indicator Genuine savings indicator		Cost–benefit analysis Risk analysis	Multi-criteria analysis Vulnerability analysis
Sustainable development	UN Sustainable Development Goal indicators Sustainable energy development indicators			Conceptual framework System dynamics Sustainability impact assessment Integrated sustainability assessment

Source: Created by authors.

All the tools and indicators of this category can be grouped into non-integrated and integrated indicators (indices). Regional flow indicators make a different class.

Environmental Pressure Indicators (EPIs), prepared by the European Union Statistical Office EUROSTAT (European Commission and Eurostat 1999, 2001), are an example of non-integrated indicators. EUROSTAT collects these indicators for EU member states and regions, in co-operation with the statistical departments of these countries. EPI consists of 60 indicators, 6 indicators for each political sphere, set by the Fifth Environmental Action Programme (Lammers and Gilbert 1999). In each political area, these six indicators may be aggregated into indices, which, in turn, make up ten environmental pressure indices. The entirety of these indicators, which consist of, e.g. damage to forests, the intensity of tourism, and polluted soil, is dedicated to assessing the environmental protection sustainability of EU member states. These indicators allow the comparison of countries and the assessment of their tendencies.

Another example of non-integrated indicators is the collection of 58 indicators, used by the United Nations Conference on Sustainable Development (UNCSD 2001). These indicators include not only the economic, environmental, and social dimensions but also the institutional dimension. Such an indicator system is not integrated. Examples may include water quality level, national literacy rate, population growth rate, gross domestic product (GDP) per capita, and a number of ratified international agreements. Since 1994, based on these indicators, countries have been preparing reports to the European Commission and accounting for their results of meeting sustainable development goals (United Nations 2002).

Sustainable Energy Development Indicators include social, economic, and environmental energy sector dimensions. The indicator system was created by the International Atomic Energy Agency, EUROSTAT, and United Nations. These indicators are not integrated and are used for the sustainability assessment of the country's energy sector. By using these indicators, it is possible to assess the sustainable energy development tendencies and take proper actions to change or promote these tendencies.

Material and Energy Flow Analysis allows analysing the resource flow structure and finding inefficient manifestations in the system. This indicator class can be used for historical reconstruction of flows and emissions as well as decision-making. Material Flow Analysis (MFA) analyses physical metabolism of the society in order to support dematerialisation processes and to reduce the negative impact on the environment, related to wasteful resource usage (Kleijn and Adding 2001). MFA studies have been performed in many countries and regions (Fischer-Kowalski and Hüttler 1998). In addition, the number of MFA has significantly increased in recent decades. This indicator class is non-integrated; these indicators include only physical flows, so they analyse the environmental dimension. MFA of all economy, calculated by EUROSTAT (2001), is a standard MFA tool for the EU countries.

World Resources Institute MFA studies for the developed world countries were the initial tool to standardise MFA in the European Union. EUROSTAT has prepared guidelines for the MFA economy assessment. In the EUROSTAT guidelines, material flow indicators are divided into three categories: input, outflow, and consumption. Each category includes different levels based on whether it involves local, foreign, or hidden flows. Hidden flows are the flows that are not included in the economic system, i.e. excavation, soil erosion, etc. (Matthews et al. 2000).

Material Input Indicators show material inflow into the economy through local production and consumption. *Material Outflow Indicators* measure all material outflow back to the environment or pollutants disposed of to the environment during the production or consumption processes. *Material Consumption Indicators* measure all material consumption in the economy.

Substance Flow Analysis (SFA) involves the regional flows of certain chemical materials and environmental losses related to them (Lindqvist and von Malmborg 2004). The goal of the SFA is to reduce the pressure of certain materials to the environment. The SFA is carried out at a regional or country level to identify problematic areas. It is useful for planning and managing an environmental policy.

Energy Flow Analysis covers energy flows in the economy. It is based on the first law of thermodynamics or energy persistence, which states that the amount of energy is constant, cannot be created or destroyed, and can only transfer from one form to another. Energy analysis of a country or a region is performed by using the *input–output energy analysis,* based on Wassily Leontief's economic input–output matrix, which analyses flows among different industry branches in the economy. In the case of energy flow analysis, among industry branches or sectors, sale flows are changed to energy flows (Finnveden and Moberg 2005).

In addition, energy analysis can be performed by using *exergy and emergy analysis*. This analysis is advanced because both the quality of energy and its quantity are considered during it (Rosen and Dincer 2001; Herendeen 2004). System exergy is the maximum amount of mechanical work that can be generated. Exergy analysis shows the efficiency of material usage and where losses form as well as where technological upgrades can be made in order to increase the efficiency of energy. The results of regional exergy analysis, conducted in Sweden, Japan, and the United States, allowed preparing the methodology of regional emergy analysis, the base of which is constituted of the expression of all resources and goods in a single unit of measurement – solar emjoules, i.e. the amount of solar energy needed to produce them.

There are many efforts to integrate sustainable development indicators in order to create one index that reflects the achievements of sustainable development (Gerlagh et al. 2002). The first efforts were directed towards supplementing new indices of national report systems such as GDP and net national income, which are meant for general welfare assessment. The latter sent false signals about the level of the achieved welfare since the achievements of sustainable development such as income distribution inequality, public safety, material overexploitation, or non-evaluation of external costs had not been assessed.

A larger variety of GDP modifications were proposed because of its limitedness in assessing environmental dimensions and willing to set an adequate life quality indicator. All of them were meant for the performance assessment of sustainable development.

Sustainable National Income (SNI) is an index created in the Netherlands. The main point of this index is GDP modification that includes sustainable resource consumption into the national income report. SNI does not directly assess social factors. It is a comparison of national income, calculated by principles of sustainability, and traditionally calculated national income. The difference between these two values shows the dependency of a country on production resource consumption that exceeds sustainable resource consumption calculated by logistic growth models (Gerlagh et al. 2002).

Index of Sustainable Economic Welfare (ISEW), created by Herman Daly and John Cobb, as well as the *Genuine Progress Indicator* (GPI), created by the organisation Redefining Progress, involves economic, social, and environmental dimensions. All these indicators are GDP modifications by adjusting national account modifications in order to include a larger number of welfare determinants such as military expenditure, environmental degradation, and natural capital depreciation. They have been calculated for many countries.

Genuine Savings Indicator (GSI) is meant for national-level sustainability assessment. This indicator is applied in the studies and reviews of the World Bank. It includes resource decay and environmental degradation indicators as well as technological changes, human resources, the export of depleting natural resources, resource discovery, and critical natural capital. Economic and environmental components are emphasised the most, but it also covers investments and education. The positive indicator value shows positive movement towards sustainability, while the negative indicator presents the movement in the opposite direction. Its advantage is that it gives a clear signal to a country about the direction of its sustainability.

Ecological Footprint is an indicator which rates resource consumption and waste formation for a certain area of land and is calculated on a country or regional scale. Ecological footprint calculation consists of a few steps. First, the average yearly food, living space, transport, product, and service consumption per capita are calculated. The land plot required to make each of the consumption needs is calculated after that, and its environmental impact based on the needed land plot is assessed. Then, after adding those land plots, a land plot required to fulfil yearly needs per capita is derived. It has been calculated for many countries and regions and is dedicated to assess the sustainability of the country; however, it can also be applied to a city or a region.

Aggregated indices are also calculated. *Welfare Index* (WI) was used in order to rate the progress of Parties of the Johannesburg Earth Summit 2002 in their sustainable development. It was calculated for 180 countries. This index consists of two indices: the *Human Welfare Index* (HWI) and the *Ecosystem Welfare Index* (EWI), which includes more than 60 aggregated indicators. HWI involves human and health parameters, wealth indicators as well as indicators of knowledge, culture, community, and equality. EWI involves dimensions of the earth, water, and atmosphere, indicators of biodiversity, and resource consumption. When they are connected to WI, these indices are given equal weight. The Barometer of Sustainability is applied for the procedure of connecting these indicators to WI.

Environmental Sustainability Index was formed in order to rate the progress achieved in sustainable development. It consists of 68 indicators that involve five different categories: the condition of environmental systems (air, water, soil, ecosystems, etc.), reduction of stress on environmental systems, reduction of human vulnerability due to environmental changes, social and institutional abilities to deal with environmental challenges, and strengthening of international standards and requirements. Even though this index assesses environmental sustainability, it also involves social and institutional dimensions. Its goal is to compare countries based on their ability to make environmental decisions.

Since 1975, the *United Nations Development Programme (UNDP)* prepares annual global reports on social human development. *Human Development Index (HDI)* of 175 countries is calculated in the report; according to it, Lithuania takes the 45th rank. HDI is a composite unit of human development.

HDI *measures the average achievements of a country based on three key components of social human development*:

- long and healthy life, which is testified by the future average life expectancy;
- knowledge and education, as measured by the adult literacy rate (two-thirds of the component weight), and the combined coefficient of those who are trying to get primary, secondary, and tertiary education (one-third of the component weight); and
- good standard of living, as indicated by the GDP.

Before calculating HDI, the indicator of each aforementioned component has to be determined, and then the HDI is calculated, which is a simple average of the three component indicators. The first UNDP report on human development was presented in 1990 for the sole goal of placing humans in the centre of the developmental process when there are economic debates, the policy is being formed, and a propagating work is performed. Each report is dedicated to the significantly relevant development discussion topic and presents an innovative analysis together with policy recommendations.

In order to make decisions on a global, country, or regional scale, policy measure analysis methods are applied, whereas local assessment methods are applied on a project scale.

In the context of sustainability assessment, integrated methods of assessment are mostly *ex ante* methods that are applied in the form of possible scenario analysis. Many of these integrated assessment means rely on system analysis and involve aspects of nature and society (Gough et al. 1998). Integrated assessments consist of many different means. Integrated environmental problem assessments are performed by using such means as multi-criteria analysis, risk analysis, vulnerability analysis, and cost/benefit analysis.

2.4.2 Conceptual Framework and System Dynamics

Conceptual framework analyses quantitative (causative) relations and applies flow diagrams, flow maps, etc. A conceptual framework can be applied in order to visualise and determine changes in the system, which positively impact sustainability, or to apply powerful computerised models for relation conceptualisation. System dynamics is the creation of computerised models to depict complex systems and experiment with them as well as monitor the long-term operation of these models and analyse various possible scenarios. General and partial balance models are an example of models applied for sustainability assessment: GEMINI, RAINS, TIMES, BALANCE, etc. IMAGE model is meant for monitoring and analysing social, biosphere, and climate system dynamics.

2.4.3 Multi-Criteria Analysis

Multi-criteria analysis is applied to assess the impact of a project or policy means based on opposing criteria. Multi-criteria analysis sets targets and tasks and seeks to weigh them and set an optimal policy tool based on all targets and tasks. This method allows rating both quantitative and qualitative data. This methodology was used to select the policy for energy and environment (Greening and Bernow 2004).

2.4.4 Vulnerability Analysis

Vulnerability analysis assesses the vulnerability of human–natural system in order to determine how sensitive a system is to changes and how it is able to deal with those changes. If it is determined that a human system or a natural system is vulnerable, a risk analysis is performed. Vulnerability analysis was performed on society and ecosystems while studying the impact of climate change (Kann and Weyant 2000).

Having analysed the sustainability assessment methods based on their ability to integrate nature and society and involve long-term periods and different area levels, it can be stated that only some of them integrate nature and society or all three dimensions of sustainability. As seen from the provided grouping of methods, only integrated methods and methods of sustainable development involve all sustainable development dimensions. Many methods involve only the environmental dimensions, especially on a product scale. Only the life-cycle costs assessment involves economic and environmental costs of the product.

In order to expand the analysis, it is aimed to integrate or join sustainable development assessment methods. For example, *life-cycle assessment* (environmental impact assessment method) was joined with *life-cycle costs assessment* (economic method) and *social life-cycle assessment*. Even though many methods involve the national level, they can also be applied at lower levels. Product level sustainability assessment methods are not related to the location; however, the efforts are made to improve these means, in order to bind them with the influence in a particular area. Forecasting methods are useful to determine the impact on sustainability on a long-term perspective. Sustainability assessment methods should be more standardised and provide more transparent results because the abundance and variety of the latter provide much confusion when rating the sustainability of policy means, projects, or products and choosing from alternative options.

In order to select proper climate change mitigation means in the energy sector, it is very important that they would be able to ensure other energy policy goals as well. Sustainability assessment of climate change mitigation means or the impact of climate change mitigation policy means on the assessment of sustainable development indicators plays an important role here.

For the purpose of comparing climate change mitigation means, the integrated indicator based on the sustainable energy development indicator system could be calculated; however, multi-criteria decision-making models allow maintaining the principle of interconnectedness among sustainable energy development indicator systems, while the calculation of integrated sustainable development indicators is the sum of the deviation of sustainable energy development indicators from the average.

3

Globalisation and Climate Change

3.1 Globalisation

The term "globalisation" can be found in Merriam-Webster's English Dictionary in 1951. This term was an attempt to express the accumulated experience in the field of education. The concept of globalisation has been rarely used in works of economists and sociologists since the 1960s; nonetheless, according to linguists, this term became popular in the scientific literature only in 1983. It is associated with T. Levitt, who published his article that year in the Harvard Business Review in which globalisation is understood as the merger of markets for individual products produced by large multinational corporations.

The processes of globalisation, the increasing interdependence of national economies, and the integrity of the global economic system, according to Baumilienė (2001), were determined by the following factors:

- Internationalisation of manufacturing and its qualitative changes
- Internationalisation of the capital market
- Liberalisation of international economic policy
- Qualitative and quantitative changes in the integration process
- Increasing role of international economic organisations

First of all, the process of globalisation was led by the internationalisation of modern manufacturing determined by the evolution of manufacturing capacities. In the course of their evolution, with the advent of new technologies and more sophisticated production, it has become apparent that manufacturing all types of production is not beneficial in any country. Due to the increasing international division of labour, there is a transition towards manufacturing specialisation. New means of transport and communication, in particular the extensive internet network, and systems allowing the management of factories and offices located in different countries from one centre, also play an important role in the development of links between manufacturers in different countries.

Another factor in the globalisation of the world economy is the internationalisation of capital, which deepens the internationalisation of manufacturing: owners of international capital are developing the specialisation of their companies located in different parts of the world and expanding the global trade of their production. At the same time, international ownership of manufacturing capital is developing.

A significant factor in the globalisation of the global economy is the liberalisation of international economic policy, which facilitates the movement of goods and capital between countries. For example, the global agreement on trade and tariffs, later the World Trade Organization, allowed to reduce duties among developed countries.

Qualitative and quantitative changes in integration play an important role in the globalisation of the economy. Integration is most often regional groupings. The emergence of a union such as the European Union has created the conditions to seek for further manufacturing progress within it creating more receptive markets for goods, capital, and labour. According to many scientists, integration is part of globalisation as it enhances the inclusion of integrated countries into a solid global economy through expanding markets.

The most advanced technologies allow competition in areas where previously it was possible to maintain monopolies within states or regions. The accumulation of innovation replaces the production of goods with the creation of services and gives flexibility to traditional businesses (unlike the first condition, the second one requires a certain state regulatory framework to have a real impact on the state economy). Strategic alliances and transnationalisation enable the possibility to combine the expansion of manufacturing into various geographical regions with the flexibility and application of specific knowledge. All this creates a new international division of labour (Segbers and Imbusch 2000).

In the world, the economies of the countries are rapidly integrating into the global economy. This process is driven by the forces effecting globalisation.

The following picture provides the key forces effecting globalisation and their target areas (see Figure 3.1).

Technological forces:

- transport revolution
- information and communication revolution

Economic forces:

- increasing incomes
- global trade
- world financial markets
- market forces

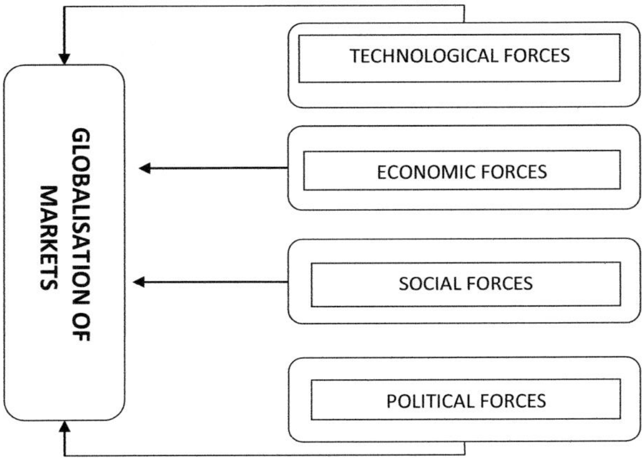

FIGURE 3.1
Key factors effecting globalisation.

Source: Brooks et al. 2010.

- global competition
- deindustrialisation in North America and Western Europe

Social forces:

- consumerism
- convergence in customer tastes
- education and skills

Political forces:

- reduced trade barriers
- intellectual property rights
- privatisation
- development of trade blocks
- technical standards, regional trading blocks

Covered areas: technology, social processes, politics, and economy are the main factors that create the potential for economic activity, and its entities to become more global, also, create the conditions for a global strategy and the distribution of resources at a global level.

Scientist T. Livett emphasises the influence of technology on the globalisation of markets. In his opinion, communications (for example, television), transport, travel, products, and processes stimulated demand for homogeneous products. The main assumption of Livett is that local tastes and needs will vanish in the long term if products become cheap enough. This hypothesis is based on the following aspects: identical products will be sold in all markets; the industry will be dominated by global corporations benefiting from the global economy (Levitt 1983).

Technological development has laid the foundations for globalisation, but political factors are no less important. Globalisation is inseparable from political processes, "since governments may shape or severely limit possibilities for private entrepreneurship" (Dreher et al. 2008). Political forces – decline in trade barriers, privatisation of intellectual property rights, development of trading blocks, technical standards, regional trade blocks (for example, the European Union, North American Free Trade Agreement [NAFTA], ASEAN).

Economic forces – increasing profits, global trade, world financial markets, global competition, deindustrialisation in North America and Western Europe. More product and service exchanges are emerging between national economies and global markets, and there is a trading system in which trade relations between any two countries may affect trade relations between the other countries. Global finance consists of credit (loans and bonds), investment (foreign direct investment [FDI], acquisition of shares and stocks), and cash (currency exchange) flows.

Globalisation is also affected by social forces such as consumerism in individual societies, convergence in customer tastes, and education and skills. The expansion of information leads to the unification of people' lives, i.e. culture, language, educational programmes, and consumer tastes. Consumerism and globalisation are related due to the following factors: access to resources and markets worldwide; manufacturing to all consumers worldwide by offering a wide range of goods, which was not easy to achieve before. Also, consumerism is associated with the essential ideas of the modern world, globalisation.

It might be also noted the international organisation UNESCO (2005), which states that UNESCO's mandate in the field of culture undertakes active action in the face of globalisation by promoting cultural diversity and intercultural dialogue. The social, educational, and cultural policies of national governments respond to UNESCO's priorities: ethical norms and principles in science and technology, human rights, the concept of lifelong learning approved at the Dakar Forum, knowledge society, peace, democracy, tolerance, philosophy, and political science value to society.

After the Second World War, the growth of globalisation accelerated, and the negative effect of international economic integration has appeared. The creation of various institutions, which try to stimulate growth and manage the negative effects of globalisation, has begun. Organisations such as the International Bank for Reconstruction and Development (World Bank) and the International Monetary Fund are created to regulate the international monetary system. To remove obstacles to international trade, the General Agreement on Tariffs and Trade was signed. Many bilateral and multilateral agreements have been signed, such as the Maastricht Treaty and NAFTA, which aim at reducing customs duties and remove obstacles to trade.

Increased opportunity to access, recycling and consumption of natural resources, and ecological and natural resource scarcity issues (greenhouse effect, lack of freshwater, waste), perceived as national for a long time, have moved to a global level, realising that natural pollution is gaining global importance (Gylys 2008). As national powers are no longer sufficient to deal with them, global governmental and non-governmental organisations are being established to deal with environmental issues.

There are certain factors that accelerate and expand the scope of globalisation:

- Large international organisations willing to operate and, where possible, control the political processes of the world (e.g. United Nations);
- International pacts – large-scale treaties covering many countries (such as the North Atlantic Treaty Organisation (NATO));
- Large and multi-member transnational economic associations (e.g. European Union, North American Association, Organisation of the Black Sea Economic Cooperation);
- International companies operating in more than one country, business, manufacturing and shopping centres, banks and corporations; and
- Certain socio-political movements called as international trends that have a profound impact on people, countries, and probably continents, such as ethnic nationalism, fundamentalism, feminism, and ecological movements.

In the time of constant changes and such factors, international organisations and integrations have emerged. These institutions and treaties are very important as they control and steer the process of globalisation in a certain direction.

The levels of impact of globalisation may be the following: global, country-specific, branch-specific, or company-specific (Figure 3.2). The global level is the growth of economic interdependence between countries and regions, the intertwining of their agriculture complexes and economic systems. The country-specific level means the increase in the openness of the economy and its integration into the global economy system. The branch-specific level: the strengthening of the interdependent competitiveness of a company within a branch in a given country, which is also competitive in other countries. Company-specific level: the development of global corporations, which is based on international manufacturing, focused on global sources of supply and manufacturing outlets.

Globalisation processes operate at different levels. Many problems cannot be assessed or investigated solely within the national state borders. One of the major factors of globalisation is economic, which emerges from the creation of multinational companies, operates simultaneously in all countries, and changes the distribution of industries, the country level, and the global economy in general.

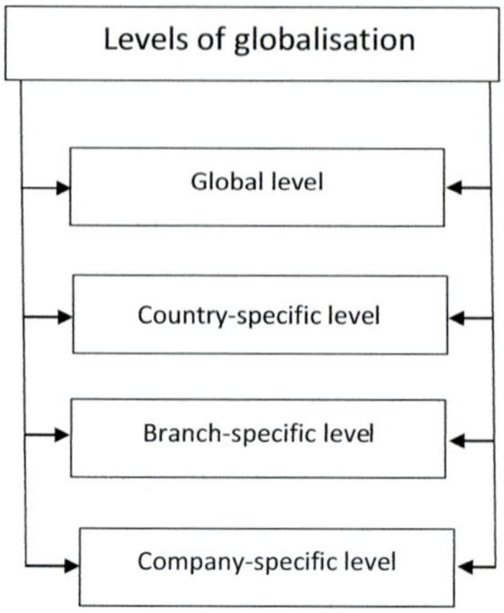

FIGURE 3.2
Levels of globalisation.

Source: Created by authors.

Already in 1988, world scientists and, most importantly, national governments spoke about the reduction of climate change effects, more specifically, greenhouse gas emissions. From that year on, the age of globalisation or the era of globalisation has begun. At the same time, a treaty was signed that established a trade relationship between Canada and the United States, which later, with the inclusion of Mexico, became the NAFTA. This and most of the subsequent agreements, events, and measures were intended to penetrate into the framework of global politics, which gives tremendous freedom to transnational corporations. Within existing agreements, the latter may produce goods at the lowest possible cost and realise them on the international market almost unhindered. For many years, it has been taught that meeting the needs of transnational corporations would boost economic growth, which would be so important to us all, and our lives would only improve.

The main features of a passing but still new era of globalisation are (2):

- privatisation of the public sector,
- deregulation of corporations' sector, and
- lower taxes of corporations that are usually paid by deducting public expenses.

3.2 Liberalisation of Markets and Climate Change

The key point that has emerged over the last three decades is that market laws and market logic have portrayed climate change, including the

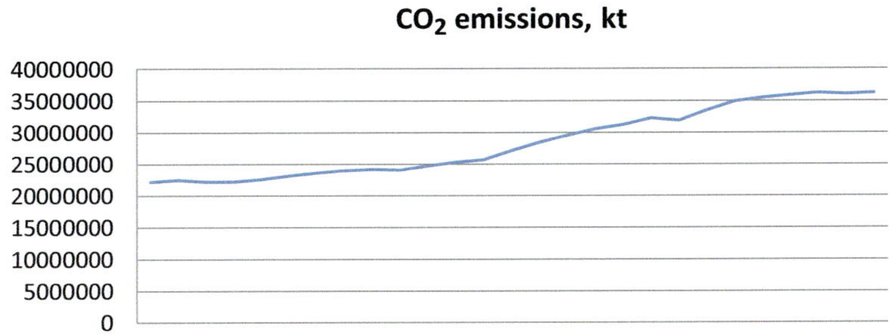

FIGURE 3.3
Gross CO_2 emissions during 1990–2016.

Source: https://ourworldindata.org/co2-and-other-greenhouse-gas-emissions.

greenhouse effect, as political heresy. Thereby, global politics successfully contributed to the most important and fundamental global warming cause – the ever-growing greenhouse effect, which is emitting more and more gas into the atmosphere. As Figure 3.3 shows, in the ninth decade of the 20th century, global pollution increased by an average of 1% per year. Since 2000, it increased by 3.4% through the entire decade. This was influenced by the successful integration of "emerging markets" into the world economy.

Pollution decreased only during the global financial crisis. However, it is worth noting that the largest contributors to the increase of pollution are the economies of the world's "emerging markets"; however, the effects of climate change hit first and the hardest the countries that emit greenhouse gases the least, countries that are disadvantaged by their geographical position and extremely vulnerable due to the severe survival conditions and poverty. Figure 3.4 shows the evolution of gross CO_2 emissions between 1990 and 2014 in the fastest developing countries.

During the globalisation period, two features of this era emerged that have played a key role in increasing atmospheric pollution, thus contributing to the adverse effects of climate change. Firstly, it is the mass export of goods over long distances. This process, which is constantly growing conditioned by globalisation, requires enormous resources of fossil fuels. Secondly, it is a model of excessive consumption created by wealthy countries and imposed on every region of the world. When combining these two elements, it becomes clear that the liberalisation of global markets has also led to the liberalisation of fossil fuel extraction, which has a particularly negative impact on climate change. This means that the free market is not able to stabilise the situation since drastic pollution reductions are only possible in times of economic downturn or the major depression.

It is becoming evident that there is a fierce battle between the human-made economic system and our planet's system, which is engaged in a constant war with life on earth. In the face of this contradiction, two mutually incompatible theories emerge: to avoid an ecological catastrophe, humanity must use as little natural resources as possible; nevertheless, to prevent the

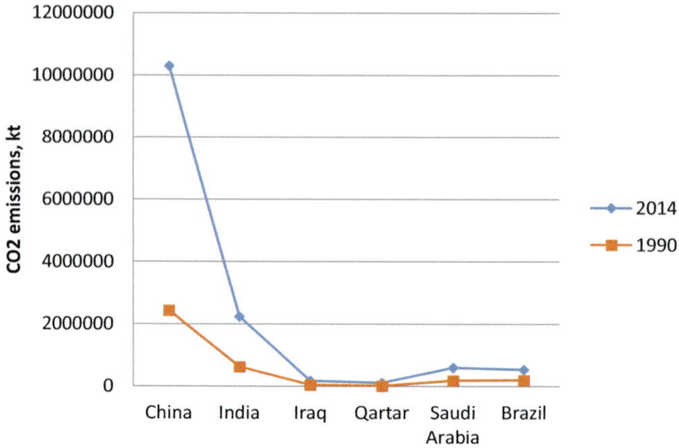

FIGURE 3.4
CO_2 emissions in the fastest developing countries during 1990–2014.

Source: Data from https://ourworldindata.org/co2-and-other-greenhouse-gas-emissions.

economic system from collapsing, unbridled and uncontrolled development is needed. Yet the concept of reducing the consumption of natural resources may be accepted by our ideology, it rejects the model of economic planning and management, where low-emission sectors of the economy are expanding and high-polluting ones are stepping back. Meanwhile, by artificially reducing the number of efficient sectors of economic activity (e.g. in case of bankruptcy), the most vulnerable groups of society may be (and are) affected.

The globalisation of the world economy is one of the most important signs of a modern economy. This concept should be implemented by reinforcing the interconnection and interdependence of its realms and participants, individual countries and regions, factors accelerating the internationalisation of the economic life, and the development of links between the various economic systems of the world, with their positive and negative effects and their impact on the world economy.

The processes of globalisation are becoming increasingly prevalent due to the higher international division of labour and cooperation in the sectors of manufacturing and services, which has provided additional competitive advantages for the well-developed countries and widened the gap between their level of development and that of developing countries.

Also, the rate of global unemployment is dependent on globalisation processes. Figure 3.5 shows the unemployment rate in regions in 2018.

The growing interdependence of world economies has given individual countries comparative competitive advantages, and globalisation became the major cause of changes in the structure of world commodities, creating the conditions for the Chinese and Singaporean economies to play an active role in the system of world leaders.

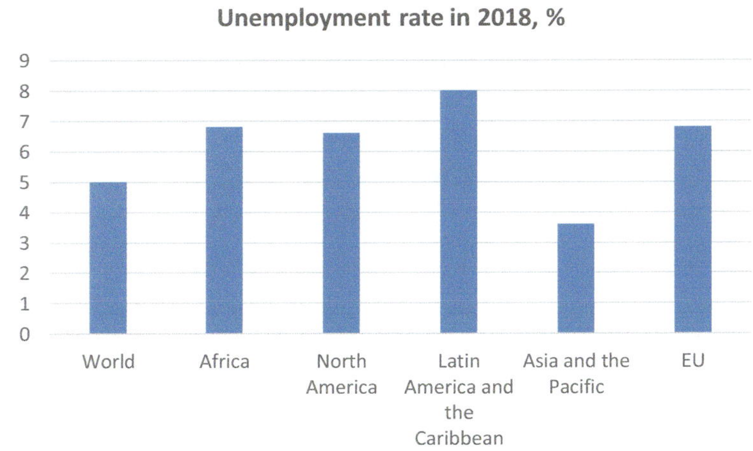

FIGURE 3.5
Unemployment in regions.

Source: https://www.ilo.org/global/lang--en/index.htm.

As seen in Figure 3.6, the working-age population has been declining globally every year since 2010. In the context of globalisation, this situation does not seem positive and puts the perspective of an ageing population into evidence, as confirmed by the data provided in the World Bank database (Figure 3.6).

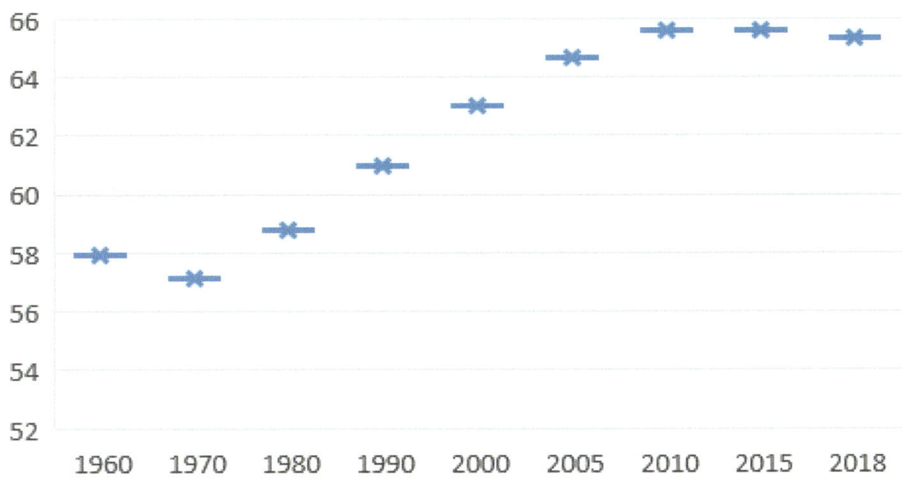

FIGURE 3.6
Population aged 15–64 (% of the total population).

Source: https://data.worldbank.org/indicator/SP.POP.1564.TO.ZS.

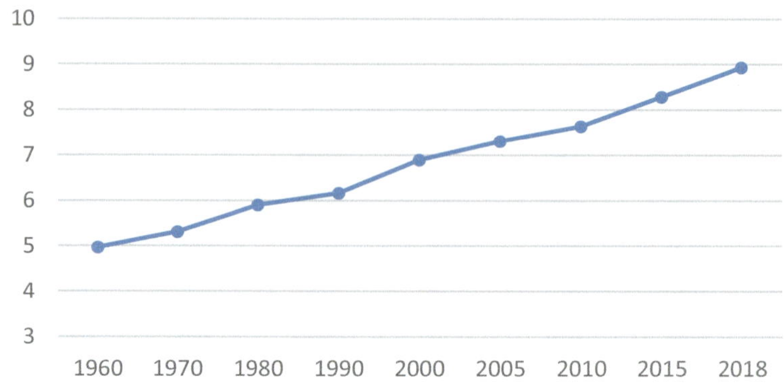

FIGURE 3.7
The amount of population over 65 years old (% of the total population).

Source: https://data.worldbank.org/indicator/SP.POP.65UP.TO.ZS.

The evolution of the population over 65 years old is shown in Figure 3.7.

The openness of national markets and the turnover of goods and the capital and labour without any significant restrictions have become one of the key features of globalisation of the world economy, though, at present, it does prevent from ensuring the economic security of individual countries and regions (the United States, the European Union, Israel, Syria, Turkey, Egypt, and others) and peace and security of the entire planet. For example, the North American Free Trade Union initially had the largest impact on the Union and Canada; however, over the last 20 years, the Mexican economy has grown significantly with its workforce threatening the economic security of the United States due to excessive and rapid migration processes.

The effects of climate change and its impact in the context of the globalisation era should be seen as a sign, which needs to be taken into consideration. Modern, rapid globalisation has created a complex global network of supply and added value chains for goods and services. The manufacturing of a single element may include full or partial equipment from all around the world. As a result, a loss caused due to climate change in one sector can cause further losses in other countries through the supply chain. Moreover, as the number of climate-change-related disasters increases, so does the aforementioned process. Figure 3.8 illustrates the evolution of the number of natural disasters recorded since 1960.

Natural disasters and consequences related to it can also be converted into financial losses, as shown in Figure 3.9. The costs include losses caused by drought, floods, extreme weather conditions and temperatures, landslides, dry mass movements, wildfires, volcanic activity, and earthquakes.

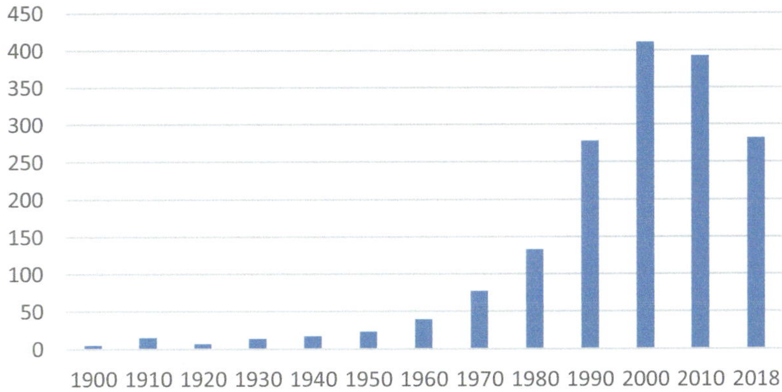

FIGURE 3.8
The number of recorded natural disasters.

Source: https://ourworldindata.org/natural-disasters.

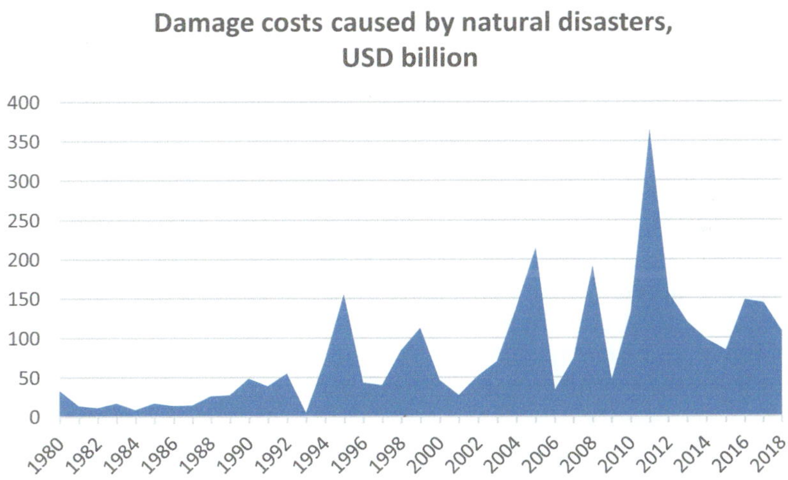

FIGURE 3.9
Damage costs caused by natural disasters.

Source: EM-DAT 2020 https://www.emdat.be/.

The negative effects of climate change depend not only on the warming of our planet but also on the resilience of our society and the global economy. Until now, population growth and models of economic development have been more important than climate change or cyclical changes in the atmosphere.

Economists have long recognised the benefits of international trade. The scale of international trade has changed over the centuries; nevertheless, it has been and continues to be a powerful force of integration, bringing together remote parts of the world and different civilisations, contributing to spreading knowledge, ideas, and information, which results in new regions and new cultures. With the rapid reduction of transport and communication costs, this trend accelerated in the 19th century. At the beginning of the 20th century, it reached unprecedented levels. Still, after two world wars and the Great Depression of 1929, there has been a noticeable increase in protectionism around the world.

Developed countries began to abandon protectionism after World War II, and since the 1970s, the developing countries have followed this example. In the 1990s, trade reforms expanded or were consolidated in South Asia, East Asia, Latin America, and Eastern Europe and, to a lesser extent, Africa and the Middle East.

Despite positive results and high expectations, disputes about the economic and social consequences of trade liberalisation remain relevant even today. Over the historical period, the following questions have been raised: Why did the trade liberalisation of some countries end in cancellation, while the others were led to prosperity, granted opportunities, and economic diversification? Does trade liberalisation promote economic growth or does this growth increase the "openness" of the economy only? Do the trade liberalisation increase country's vulnerability and poverty? Or does it help to reduce poverty? Contemporary scientists are given the responsibility to answer these and many other questions.

The liberalisation of international trade has been a key element of successful growth strategies. However, the application of free trade conditions has had different consequences in various regions of the world. Some countries, having opened their markets to free trade, have felt strong positive economic changes, and the reduction of trade barriers has helped to boost productivity and growth. Nevertheless, at the same time, other countries did not have the positive market liberalisation impact as the reduction of trade barriers is only one of the tools that is necessary for the countries to expand their participation in international trade and benefit from it. Another important point, which explains different open markets or free trade, is that international trade is a possibility, not a guarantee (Bineswaree and Freund 2004). The experience gained over the last few decades shows that attempts are being made to adjust market failures that have become extremely expensive.

As shown in Figure 3.10, the global trade market liberalisation has increased exports of goods from USD 305 million (in current prices) in 1970 to USD 17,880 million (in current prices) in 2017.

Figure 3.11 reveals trends in global import growth. Since the start of market liberalisation, global imports of goods have increased more than 57 times. Thus, the trade reform has increased export or the growth of economics of some countries, however, at the same time, has had extremely high social and economic adjustment costs. The experience of countries that underwent this

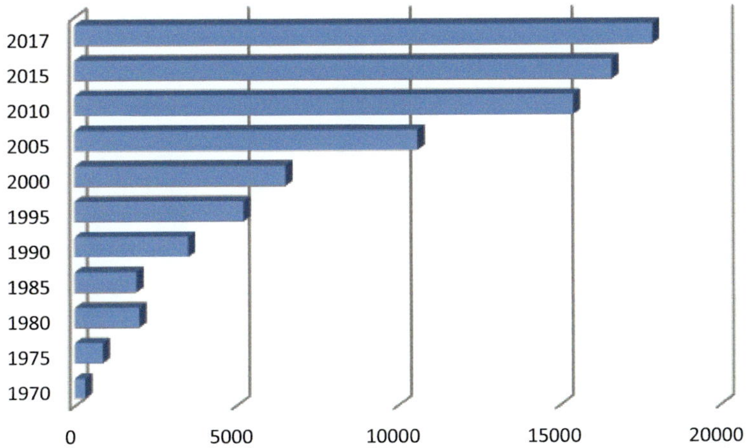

FIGURE 3.10
Global export of goods during 1970–2017, USD million (in current prices).

Source: https://data.worldbank.org/topic/trade.

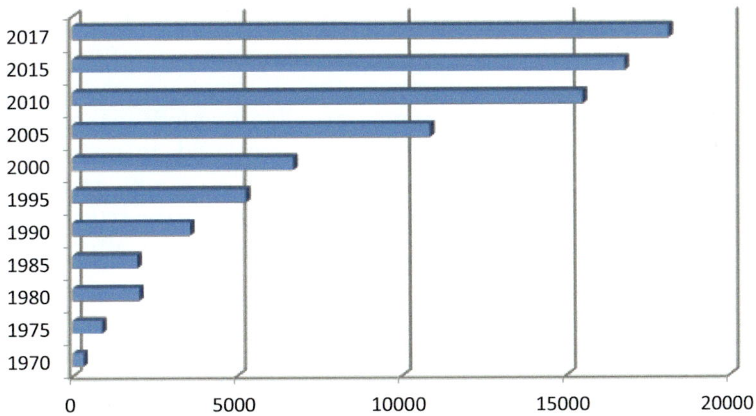

FIGURE 3.11
Global import of goods during 1970–2017, USD million (in current prices).

Source: https://data.worldbank.org/topic/trade.

economic growth does not deny the benefits of the free market but clearly demonstrates that any trade reform must be carried out deliberately as a part of an effective strategy of growth.

During 1980–2000, the global integration of goods, services, and investment flows has increased heavily. The largest increase in investment flows

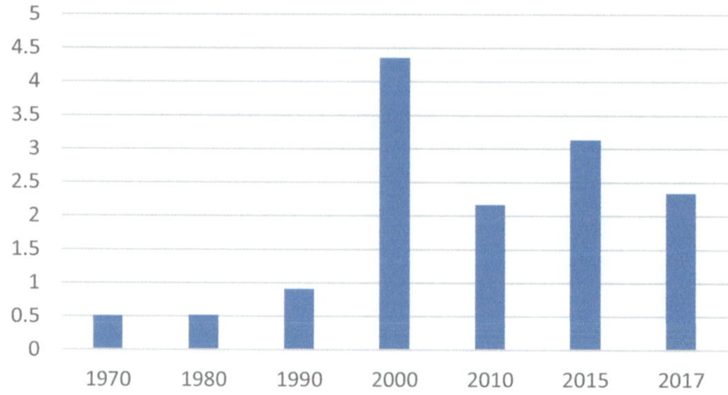

FIGURE 3.12
The evolution of foreign direct investment.

Source: https://ourworldindata.org/trade-and-globalization.

was observed as FDI. In Figure 3.12, the evolution of FDI from 1970 to 2017 is presented.

Developed countries experienced a much greater increase in FDI flows that more than doubled their gross domestic product (GDP), while developing countries experienced more modest growth (see Figure 3.13).

However, while trade integration can enhance an effective growth strategy of a country, it cannot ensure the effectiveness of its implementation. Other complementary elements are also needed; for example, reliable macroeconomic management for trade-related infrastructure and free trade-related institutions throughout the economy, proportionally large or sufficient investment in human capital and infrastructure, or even establishing strong institutions; since such reforms are often difficult to

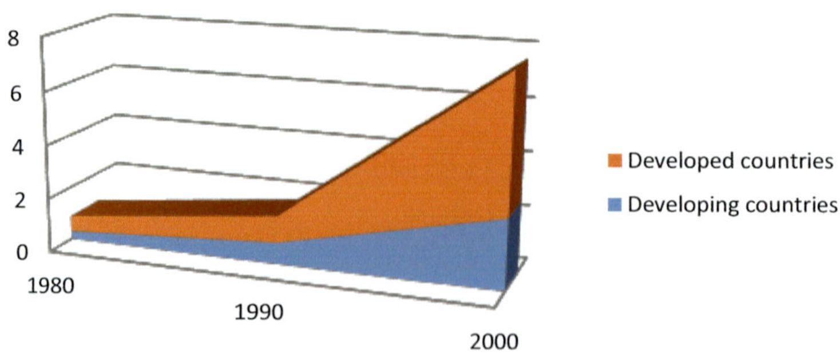

FIGURE 3.13
Share of FDI in developed and developing countries. FDI, foreign direct investment.

Source: https://data.worldbank.org/products/wdi.)

implement and the focus is often on trade policy rather than on the overall strategy for growth. Without a proper solution to the problems of free trade and a proper institutional framework for a market economy, it is difficult or often impossible to successfully implement the overall economic growth strategy of a country.

Due to the liberalisation of markets, as an economy accumulates physical and human capital, shifts its comparative advantage towards more capital-intensive activities, and becomes internationally competitive on a wider range of goods and services, it will inevitably accelerate the trade. But is higher trade the result or the cause of its growth? Most likely it is both a cause and a consequence (Tang and Harrison 2004).

There are many possible ways to open an economy. Policymakers face the challenge to determine which is the way best suited to their political economy, institutional constraints, and initial conditions (Tang and Harrison 2004). Since these vary from country to country, it is not surprising that countries that have successfully promoted growth through trade liberalisation have followed a variety of policy approaches. The experience of different countries with regard to differences in the timing and pace of reforms is evident. Different countries open different sectors at different speeds (e.g., Bangladesh and India); other countries have achieved partial liberalisation by defining export areas (e.g., China and Mauritius), while others are unilaterally engaged in trade reforms (e.g., Estonia) by participating in regional trade agreements (e.g., Mexico and Central and Eastern European countries that have now become the members of the European Union).

The effect of trade liberalisation is not always positive. Supposedly, trade reforms are expected to have negative consequences in the future, as most societies sustain a relatively complex relationship between the rich and the poor. The wealthy classes have been more successful in exploiting free trade conditions for their own benefit. It is naive to expect for trade reforms to increase the income of the unskilled, although according to data from the 1990s, the relationship between trade reforms and poverty is still indirect. Even in cases when trade policy reduces poverty, the main problem with income distribution is that countries need to help people move from the contracting (import-competing) sectors to the expanding (exporting) sectors.

The liberalisation of markets can also indirectly affect poverty (Winters et al. 2004). Several important consequences have been identified: (1) trade reforms can affect the employment opportunities and remuneration of the poor; (2) reforms also have a direct impact on the prices of goods and services that consumers typically pay for; (3) trade reforms can reduce government revenue and, in turn, reduce social spending, which is particularly important for the poor (various social allowances, including several benefits); and (4) market liberalisation can increase income volatility, leading to job losses or lower-skilled or lower-paid jobs. Even with the general decline in poverty, which is often reported in statistics (see Figure 3.14), the liberalisation of markets and steady population growth (Figure 3.15) may result in

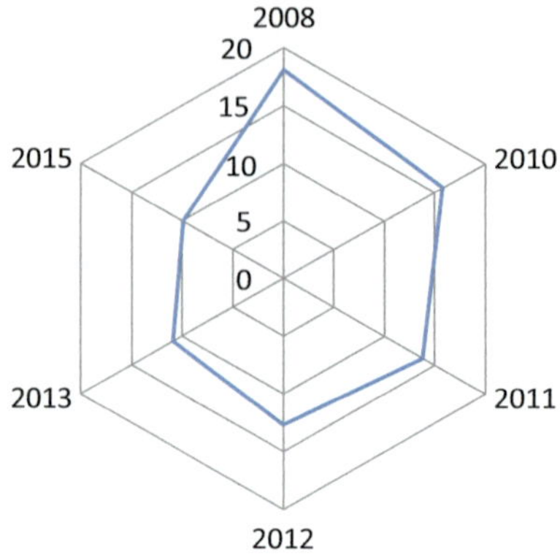

FIGURE 3.14
Residents living on less than $1.9 a day, according to the purchasing power parity of 2011.

Source: World Bank 2020.

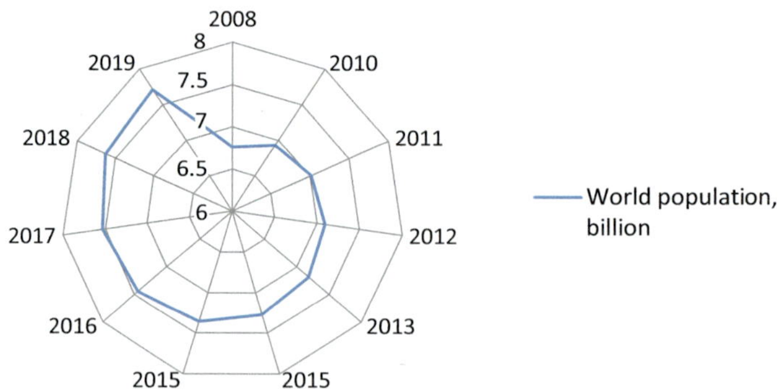

FIGURE 3.15
The evolution of the world population.

Source: World Bank 2020.

many households losing or significantly reducing their income (Tang and Harrison 2004).

An empirical study of the impact of trade liberalisation did not persuade some sceptics who argued that trade liberalisation did not bring significant or expected economic benefits to countries, and some even concluded that trade reforms led to economic collapse and "deindustrialisation" (Salinas and Aksoy 2006).

Studies conducted by Copeland and Taylor (1994) have clearly shown that relatively low incomes could give a country a comparative advantage in pollution-intensive goods, even if the costs of environmental damage are optimally internalised. It is clear then that free trade increases revenue, which results in scale and technical effects that outweigh each other; thus, the net impact on the environment depends on the composition of the impact. Therefore, trade liberalisation promotes an increase in environmental damage in poorer countries.

The liberalisation of environmental goods using recycled or green products and renewable electricity could help mitigate climate change by removing tariffs and non-tariff barriers in the market. Trade liberalisation is only one of many factors, including GDP, FDI, environmental regulatory frameworks, and technical assistance, which affects the actual trade in climate mitigation goods and their dissemination. Fiscal incentives, investment systems, and intellectual property-related costs also determine access to and availability of climate change mitigation technologies. However, many developing countries have other goals, such as the protection of sensitive industries and the development of domestic capacity, which may discourage them from fully liberalising climate change mitigation goods.

3.3 The Impact of Climate Change Mitigation Policies on Business Development

Business development requires new processes to create and deploy growth opportunities between organisations and within the organisation itself. Business development is the creation of long-term value of an organisation from customers, markets, and relationships. Business development can also be any activity of a small or large organisation, non-profit or for-profit company, which in some way "improves" the business. In today's society, business development must be sustainable so as not to harm both the environment and society. Therefore, it can be argued that business development and climate change, today's global issue, are integral components of sustainable business growth. Kauggman and Tebar (2009) argue that business attitudes to climate change are driven by a variety of factors, including government policies and regulation by consumers and other stakeholders.

In recent years, governments in OECD countries have implemented measures to reduce domestic climate policy frameworks, including greenhouse gas emissions. World leaders of the UN recognise that climate change is a problem that requires urgent and decisive actions from government, business, and companies. Business leaders commit to taking practical steps to increase the use of energy and efficiency. In practice, business decisions are driven by government policy, change, consumer demand, and technological innovation.

Sustainable business, from a risk perspective, is a good way to respond to potential impacts on climate. It is generally acknowledged that the Earth's climate is changing and greenhouse gas emissions, caused by human activity–business activity, are increasing. Most large companies currently see managing their greenhouse gas emissions and reducing energy consumption as an integral part of their environmental management practices. Many have set targets to reduce their greenhouse gas emissions, in cooperation with their suppliers and customers to reduce emissions, and encouraged governments to adopt policies aiming at reducing global greenhouse gas emissions. According to Sullivan (2013), most companies have focused their efforts on those actions that provide clear and relatively low returns and those with relatively low costs. As a result, several business strategies of companies are singled out, which would have relatively less impact on climate change. First of all, companies need to rethink their capital investment processes.

For many companies, the single biggest opportunity in the future to prove their business and create long-term business value is when they invest capital in new projects or new products or update existing equipment. If in the process of those decisions climate change is taken into account (e.g. by applying shadow carbon price considering several scenarios of regulation and the physical effects of the climate change), this maximises the likelihood that companies will make decisions that do not involve excessive assets or loss of revenue due to regulatory or other actions on climate change impacts. Companies also need to develop information, knowledge, and experience on climate change. Companies, most negatively affected after the integration of climate change into their business strategies, stress how much time and effort they invest in testing new technologies and new methods. That is, when it comes to investments (e.g. in a new vehicle fleet), they fully understand not only the financial aspects of their decisions (i.e. costs and benefits of a particular technology) but also operational and other effects (e.g. new equipment, availability of specific raw materials).

Therefore, it can be stated that climate change is a strategic issue for business. The companies, which will be best prepared to respond to the inevitable business and social pressures arising from climate change, will be those which have recognised climate change as a strategic business value driving force. Table 3.1 provides opportunities and risks related to climate change for business sectors.

Meanwhile, Fontinelle (2010) in his article identified seven ways climate change affects business development and activity:

1. First of all, emission control systems require capital expenditure. Some companies may have to invest substantial amounts to upgrade polluting installations and install emission control systems to comply with increasingly stringent rules on greenhouse gas emissions. This would largely affect energy and utility companies that operate heavy polluting facilities such as refineries and power plants. According to Smith (2016), in today's society, the focus is on addressing the most important environmental problems that have arisen as a result of economic development and are of considerable concern. For this

TABLE 3.1

Opportunities and Risks Related to Climate Change for Business Sectors

	Opportunities	Risk
Food industry	1. Increase in global demand due to climate challenges (e.g. drought). 2. Longer growing season of grains, vegetables, and fruits, due to the warmer climate can improve productivity. 3. Countries that are "rich in water resources" will be able to be more competitive compared to other countries.	1. Extreme weather events can affect transportation, refrigeration of raw materials, such as milk. 2. Warmer weather may increase the number of insects carrying dangerous diseases. 3. Current production and market standards are water and energy-intensive and in the future may become more expensive.
Pharmaceuticals, medicine, new technologies	1. Growing opportunities to address global health problems caused by climate change.	1. Rising ambient air temperatures can lead to a deterioration in the quality of water, which has a negative impact on the chemical processes of water cooling. 2. Lots of chemical and pharmaceutical industries are located on the coasts, in particular in Cork Harbour, and the Shannon Estuary, which may be disrupted due to the rising sea levels and the risk of flooding.
Insurance/ financial services	1. New insurance products related to climate risk may develop. 2. The risk of investment funds may increase if capital investments are made in assets that are vulnerable due to climate change.	1. Climate change can create difficulties for global financial markets in terms of risk perception, especially with regard to insurance capacity. 2. The sensitivity of the insurance sector will be heightened and most likely to suffer losses from the effects of extreme weather.

Source: Created by authors based on Adaptation to Climate Change: Issues for Business Summary Report.

reason, companies seek to act in environmentally friendly ways or in an attempt to solve some of the problems that industry creates for the environment. Large global companies are making significant investments to reduce environmental pollution. To this end, they use renewable energy sources and implement the latest technologies in the equipment that are less harmful to nature.

2. Potential internal investment funds and trading laws. Under the programme on climate change mitigation by the European Union, which includes emission reduction provisions, a certain emissions target is imposed on companies, allowing them to legally emit only a certain amount of greenhouse gases to the air. Companies that exceed emission standards will have to purchase additional credits,

thereby inhibiting profitability. On the other hand, companies with excess credits could sell them for extra money.

3. The range of prices for services and goods. Even companies that do not cause significant pollution may be indirectly affected by climate change laws, as their suppliers and/or consumers may be affected. Price ranges are likely to be caused by large-scale costs due to increased transport expenses or higher electricity prices.
4. Evolution of weather conditions. According to Yongqiang et al. (2010), climate change which is being affected by air pollution is expected to change weather conditions worldwide. Storms are expected to become more severe and cause a number of negative consequences. This is likely to cause higher losses for insurance companies and could make ocean shipping even more dangerous. Established farming areas may become less fertile, which would lead to losses for agricultural enterprises.
5. Change of demand for goods. The combination of changing prices and changing weather patterns is likely to reduce demand. For example, if global temperature rises, demand for cold weather production, such as heating fuel, may decrease.
6. Commitments under the Kyoto Protocol, the European Union Emissions Trading System and other foreign regulations. According to Heakel (2015), it is important to bear in mind that many state-owned companies operate abroad and may be subject to various laws and regulations on climate change. For example, while many countries, such as the United States, have not signed the Kyoto agreement, they have subsidiaries in other countries that are officially trying to comply with the mandate of the protocol. The European Union has an emission credit system for large emitters. It is difficult to assess the potential impact of all this divergent regulation.
7. Changing public opinion on corporate activities. Like Fontinelle (2010) in his article singled out, so did Delventhal (2016) argue that reputation is essential for many companies. Increasingly, negative public opinion is being heard about companies whose activities are polluting the environment. Many companies today try to promote a green image and to contribute to climate change mitigation policies. Companies of this type are improving their activities to reduce environmental damage.

Sullivan (2013) has envisaged based on the scale of greenhouse gas emissions reduction that is necessary to avoid the worst effects of climate change, and due to the performance of the policy actions, a complete transformation of the global economy will be achieved. The author had in mind that not only sectors, such as electricity generation, transportation, and heavy industry, but also their activities directly contribute to environmental pollution and climate change greenhouse processes, yet the activities of almost every business, business models, and strategy will have to be fully revised. Businesses also contribute and will contribute to reducing greenhouse gas emissions through capital allocation, innovation, and the development of new technologies, as well as by influencing governments' actions on climate change. Richard et al. (2013) could also contribute to the insights of Sullivan (2013), who stated in their article that the European Union has created the world's largest market-based system for reducing greenhouse gas emissions: the Emissions Trading System, which started in 2005. The European Union has established a system that controls the activities and development of companies in such a way as to minimise the negative impact on the environment and climate change. To achieve this, the European Union has introduced emissions charges, standards for greenhouse gas emissions, and the limits that many companies have to respect. At the same time, the European Union has created a favourable space for companies that use renewable energy sources in their activities, by investing in and encouraging them.

In scientific literature, the impact of climate change mitigation policies on business development is examined from both a theoretical and a practical point of view. In particular, climate change mitigation policies are linked to the Paris Agreement, which took place in December 2015. The Paris Climate Agreement aims to prevent the most severe effects of climate change; therefore, the signatory countries agreed to limit the global mean surface temperature increase that, compared with the pre-industrial period, would stay below 2°C. To achieve this goal, the quantity of global greenhouse gas emissions should reach the highest level as soon as possible and then sharply decline. According to the Council of the European Union, the main elements of the new Paris Agreement are as follows:

- *Long-term objective*: Governments agreed to keep the increase of global average temperature well below 2°C compared with pre-industrial levels, and to make efforts for restriction on 1.5°C.
- *Contributions*: Before and during the Paris conference, countries presented comprehensive national climate action plans to reduce their emissions.
- *Ambition*: Governments agreed to report every 5 years on their contributions to set even more ambitious goals.
- *Transparency*: They also agreed to inform each other and the public of their progress in achieving their objectives to ensure transparency and supervision.
- *Solidarity*: The European Union and other developed countries will continue to provide climate finance to developing countries to help them reduce emissions and increase their resilience to the effects of climate change.

Rogel et al. (2016) conducted an analysis of the Paris climate change mitigation policy and presented the policy conclusions and their prospects and impact.

Firstly, climate change mitigation policy plans related to greenhouse gas emissions are a historic achievement. The Paris Agreement requires consistent, increasingly ambitious, nationally determined contributions that are subject to strict transparency guidelines. Even since 2015, limiting the increase of the average temperature to no more than 2°C compared with the pre-industrial levels is a societal challenge. National contributions represent a new era for climate policy under the Paris Agreement and are both an invitation and a call for further action. The authors distinguish two promising developments that are the essence of the Paris Agreement. First of all, it is becoming increasingly clear to decision-makers that measures to reduce greenhouse gases have significant socio-economic benefits. The actions of almost all countries improve the prospects for further collective action, which makes it easier to understand additional measures or to strengthen the existing ones. Secondly, the recent unprecedented involvement of non-state subjects such as businesses, citizens, and religious organisations illustrates the increased awareness and momentum of climate action. Given the high potential for reducing emissions as a result of both of these options, it will be crucial to support and authorise national and non-state actions. This insight also opens up important future studies and evaluations. The research community will have to break from a one-sided climate-policy-oriented approach and develop new concepts and systems to help achieve public objectives, including the portfolio of climate, food and energy security, public health, and other sustainable development objectives.

Meanwhile, Dimitrov (2016) has also examined the Paris Agreement and presented both positive and negative sides of politics. The key provisions include a global objective to keep the temperature increase to "well below 2°C" and "to pursue efforts to limit the temperature increase to 1.5°C" (Article 2) and the objective of achieving the global peaking of emissions "as soon as possible" (Article 4.1). "All Parties are to undertake and communicate ambitious efforts" (Article 2) and "each Party shall prepare, communicate and maintain successive nationally determined contributions that it intends to achieve" (Article 4.2), which shall be revised every 5 years. Throughout the text, the language is strictly worded ensuring progression over time. Developed countries "should continue taking the lead" with all of the absolute emission reduction goals, while developing countries are under a weaker obligation and "should continue enhancing their mitigation efforts, and are encouraged to move over time toward economy-wide emission reduction or limitation targets in light of different national circumstances" (Article 4.4). The accompanying Decision contains a provision that, once it enters into force, Parties are obliged to set a new collective financial goal "from a floor of USD 100 billion per year" (Paragraph 54). The Paris Agreement also creates a market-based sustainable development and carbon trading mechanism proposed by Brazil and strongly supported by Japan and other developed countries. Global stocktaking is to take place in 2023.

Politically, the Paris Agreement generally favours the developed Nordic countries, which have won most of the central battles. The new climate covenant meets all the key requirements of the United States and is based on a model of global climate governance that Japan proposed in the early nineties: the "pledge and review" system.

The Agreement is the least fair for the African Group and other least developed countries. It does not include references to their particular circumstances.

The advantages of the Agreement appertain to principled obligations to act, regularity and development of national policies, international transparency, and accountability. The Paris Agreement is weaker on the long-term global goal, adaptation policy, compensation for loss and damage, and technology transfer. Most importantly, the Paris Agreement lacks specificity on the international division of labour for reducing emissions. Sharing of responsibility has been a major challenge in global negotiations. The precise link between national mitigation policy "contributions" and the global policy goals is not clearly defined. The complexity and experimental nature of the Paris Agreement make its evolution difficult and have already caused controversy. In the ongoing debate among international lawyers on the character of the Paris Agreement, scholars argue that it is a legally binding treaty creating obligations and a complex combination of binding and voluntary provisions.

Under climate change mitigation policies, including the Paris Agreement, the main instruments for mitigating climate change relate to *energy efficiency and renewable energy sources.*

Štreimikienė (2010) conducted a study identifying the link between energy efficiency–security and climate change mitigation policies (Table 3.2). To create a sustainable integrated European climate and energy policy package, the most recent EU initiative identifies EU energy policy priorities and the close interplay between climate change mitigation and increasing energy security–efficiency. Integrated policy packages can deliver coordinated policies and the effect of the synergy. Policy evaluations should be based on their impact on the key EU sustainable energy policy development goals: greenhouse gas mitigation, energy efficiency improvements, the increase of the use of renewable energy sources, and the increase in the energy consumption. The external costs of climate change and energy security can be used to assess policy impacts in monetary terms.

Table 3.2 presents the synergy between climate change and energy efficiency measures based on the analysis of the EU's energy policy. As already highly discussed in the new energy policy, the European Union is committed to reducing greenhouse gas emissions at least by 40% by 2020 compared with the level of 1990, to increase energy efficiency and by 2020 achieve 10% of the biofuel component to be used in vehicles. Reduction measures for greenhouse gas emissions mainly positively affecting the energy use are illustrated in Table 3.4. The development of carbon collection and storage technologies

TABLE 3.2

Interaction of Climate Change and Energy Efficiency–Security

Energy option	Climate change impact	Energy efficiency–security impact
1. Replacement of the Energy Generation		
Promotion of renewable energy sources	Zero GHG emissions	Renewable energy sources can often be obtained domestically
Move from lower to higher carbon amount	Higher GHG emissions	Coal is a substantially secure source of energy as each country has domestic suppliers
Move from higher to lower carbon amount	Lower GHG emissions	Depends on the country
2. Change in Transport Fuel		
Increase in biofuel use	Zero GHG emissions	Statement
Increase in natural gas use	Positive	Neutral
Increasing import diversity	Depends on the country	Positive
Growing fossil fuel production in the country	Depends on the fuel type	Positive
Clean energy technologies	Positive	Positive
Carbon collection and storage technologies	Positive	Positive

Source: Created by authors based on Štreimikienė (2010).

is also an important factor in securing future energy supplies. EU carbon market development and an open and competitive internal energy market would also help to increase energy efficiency and security; however, prior to the implementation of all policies and measures, they need to be aligned with energy policy development goals (reducing greenhouse gas emissions, increasing energy efficiency, increasing use of renewable energy sources, and increasing energy security) by assessing the impact of key EU sustainability measures.

The real impact of climate change mitigation policies on the business sector includes studies on how the business sector responds to climate change mitigation policies.

Media seems not to influence the perception of the mitigation policy or the implementation of this policy. Both the United States and South Africa had higher media attention on climate change; still, both countries see economic growth as more important than environmental protection. Therefore, media attention does not provide public support for climate change mitigation policy nor does it contribute to the implementation of that policy. In Germany, meanwhile, environmental protection comparing with economic growth is more important. In this country, climate change mitigation policies are assessed and strictly adhered to.

South Africa has relatively high emissions for a developing country, mainly due to its carbon-intensive energy structure and dependence on mining and minerals processing. South Africa's climate change mitigation policy was assessed on the basis of the country's ability to target its most intense emission categories, i.e. electricity, transportation, and industry. For example, the average energy emission level of a developing country is typically 49%, while South Africa accounts for almost 80% of its total emissions. Electricity generation accounts for half of South Africa's energy emissions and only slightly less than 40% of all emissions. The share of industrial transport and energy consumption contributed just under 10%, and emissions from industrial processes accounted for about 14%. Agriculture and land-use developments in South Africa account for 5% (National Climate Change Response White Paper 2011).

In Germany, the 2014 Climate Action Programme is one of the most advanced climate mitigation policies in the world. It commits to reducing greenhouse gas emissions by at least 40% by 2020. Germany will engage in emission trading, promote to invest in higher energy, increase in regulations for reducing greenhouse gas emission and renewable energy to target its industrial sector, and thus implement climate change mitigation policies. Germany will tax fuels used for heating purposes within households and implement subsidy programmes to promote the use of renewable energy in its heat market to reduce household energy consumption.

The United States has recently adopted several mitigation policies, but all of them are selectively focused on a single emissions sector. The focus is on the United States transport sector, which is the most emissions-intensive sector. The United States also promotes the use of low-sulphur fuel, coal, and natural gas and promotes the use of market principles like emission banking and trading and, however, does not set deadlines or penalties for not utilising these programmes.

Whereas the objectives of climate change mitigation policy are closely linked to energy, Bassi et al. (2013) conducted a study providing UK business insights on energy and climate change policies (Table 3.3) and examined the impact mitigation policies have on the UK business sector.

According to Bassi et al. (2013), the views of these business sectors on the impact of climate change policy and the environment have shown that climate change policy is relatively relevant in today's business model and development opportunities, not only in the United Kingdom but also worldwide. Clearly, the relatively narrow sample of companies surveyed means that the evidence is not exhaustive, but provides useful information on business perceptions of energy and climate change policies. The authors of the study presented how a separate climate change policy affects business sectors in the United Kingdom. Table 3.4 provides indirect carbon taxation indicators for key energy and climate change policies.

UK businesses have several directions on climate change policies. Some of them represent the direct cost incurred by enterprises when consuming a certain amount of energy. There is also policy applied at the point of production ("upstream"), but their costs are passed on to "downstream" energy

TABLE 3.3

Business Insights on Energy and Climate Change Policy in the United Kingdom

Sector	Business Insights on Energy and Climate Change Policy in the United Kingdom
An energy supplier	The European Union Emissions Trading Scheme (EU ETS) had a positive impact on low-carbon investments since they were first introduced, but a major shortcoming has been the system's failure to deliver a long-term, sustainable and consistent price. "Consistent" means that "the ETS price should avoid boom and bust cycles and provide greater stability to low-carbon investments by continually reflecting the carbon price". In parallel, however, the reform of the EU ETS is considered necessary to ensure a coherent price signal and to enable the carbon market to respond to future unforeseen economic circumstances. Overall, the business believes that the UK government has been learning from experience and improving energy and climate change policies over recent years. It also believes that domestic policies would be more effective if they tackled carbon dioxide emissions at source, rather than at the point of energy use, as some of them do currently.
An electrical goods manufacturer	Half of the emissions from this business are controlled by climate change agreements. This business believes that the system of targets and tax discounts provided by climate change agreements is a useful incentive, as it effectively allows for double saving on energy bill – first through a discount on the climate change levy, and again through energy efficiency savings. Senior management can easily "sign up" to this type of incentive and therefore support the energy efficiency investment needed to ensure compliance. The tax relief offered under the enhanced capital allowance has also a positive impact on investments in enterprises of low carbon emissions. The carbon reduction commitment energy efficiency scheme is not applied to business, but the investments of large scale were initially invested in order to prove its non-eligibility. The Green Deal is not considered to be very cost-effective, given its relatively high-interest rate compared to the interest rates that the company can secure through conventional bank loans.
A building material manufacturer	The complexity of the policy is an important issue for this business. As a large and complex business with many large and small locations across the country, it depends on different climate policies that overlap and interact. Furthermore, companies consider that the administrative costs associated with compliance with the carbon reduction commitment energy efficiency scheme are particularly high. As far as climate change agreements are concerned, although they are a good incentive to invest in energy efficiency, long negotiations on the objectives set out in the agreements have resulted in relatively high administrative costs. The climate change levy, in contrast, has low administrative costs, but its impact on investment decisions has so far been relatively low, as it is considered as the part of the operating costs of the business rather than a real lever for investment. Policy uncertainty, including uncertainty about the price of the European Union Emissions Trading Scheme, is perceived as a "drag" on low-carbon investment. Policy credibility, on the other hand, can make policy much more effective. In this respect, the company considers the landfill tax to be a positive example, as its objective has remained clear and consistent over the time between governments.

(*Continued*)

TABLE 3.3 (*Continued*)

Business Insights on Energy and Climate Change Policy in the United Kingdom

Sector	Business Insights on Energy and Climate Change Policy in the United Kingdom
A retailer	About 40% of the emissions generated by a retailer are included in climate change agreements while the rest are subject to the full rates of the climate change levy. Its high emissions are also subject to the carbon reduction commitment energy efficiency scheme. It is considered to be the biggest financial and administrative burden for all energy and climate change policies. Indirect policy costs, passed on through electricity prices, are also considered significant, in particular for renewable energy. At the same time, the retailer reports that renewable energy and the feed-in tariff scheme were a positive incentive to invest in renewable energy sources. In general, the business notes that overlaps and inconsistencies between policies mean that it pays for its carbon emissions several times over. Subsidies for renewable energy sources are considered to be disproportionately high and not always well targeted. Political uncertainty is considered as a fundamental issue. Increasing the certainty of political costs and their future increase is considered to be important in order to prevent future energy prices from rising too sharply and unpredictably.

Source: Created by authors based on Bassi et al. (2013).

TABLE 3.4

Instruments of the Climate Change Mitigation Policy in the Energy Sector of the United Kingdom

Policy	Instrument	Sector	The Number of Companies	Costs/Revenues (2013)
CCL (Climate Change Levy)	Tax on energy use	All sectors (except for very low amount of energy using sectors)	900,000	EUR 600 million
CCA	Tax on energy use	For energy intensive users		2,800
CRC	Tax on carbon	Medium/large companies consuming more than 6,000 MWh/year	2,100	EUR 700 million
EU ETS	Cap and trade system	For energy-intensive users	600	EUR 300 million
RO	Subsidy to renewable resources	Energy suppliers and renewable electricity generators	2,300	EUR 2,000 million
FIT	Subsidy to renewable resources	All sectors, using renewable energy resources	11,000	EUR 150 million
GD	Loan on energy efficiency devices	All sectors	n/d	n/d
RHI	Subsidy to renewable heat	All sectors	1,500	EUR 40 million

Source: Created by authors based on Bassi et al. (2013).

users through higher electricity prices. These include the EU ETS itself (where it applies to power generators), as well as measures supporting renewable energy, for example, the Renewables Obligation (RO) and the small-scale Feed-in Tariffs (FITs). Other policy directions provide financial support to companies investing in low carbon and energy efficiency technologies, without affecting electricity costs. These include the Renewable Heat Incentive (RHI) and the Green Deal. Subsidies are also available for research, development, and deployment of low-carbon technologies. These policy directions vary widely in terms of their structure, impact and type, the implicit carbon tax they levy on used energy, and the overall costs.

Not only foreign but also Lithuanian scientists have delved into climate change mitigation policies and its impact on business sectors. Baležentis et al. (2016) conducted a study to examine whether the Lithuanian economy is achieving the stated objectives of the European Union's climate change mitigation policy. To determine the environmental performance of selected economic sectors in Lithuania, the World Input–Output Database was used.

The results showed that between the year 1995 and 2009, energy consumption and carbon dioxide emissions decreased by approximately 12%. The key engine for energy consumption and emissions was a decrease in energy intensity (a decrease of more than 50%), while the carbon factor decreased by 12%. Therefore, the energy combination in the analysed economic sectors can be further improved. In fact, only oil production showed an increase in energy intensity. The researchers also established the Environmental Performance Index (EPI). Analysed economic sectors were divided into four groups based on the average EPI and its growth rate. Sectors of pulp, paper, and agriculture were indicated as the best performing group with the highest average EPI and the highest growth rate. Sectors of oil extraction and air transport were distinguished as the ones with the lowest environmental profile. The latter sectors should, therefore, be of particular importance in the development of sustainability strategies. According to the authors, the fuel mixture change of economic activity may be the most important factor in the air transport sector's environmental performance and clean technologies in the oil sector sustainability improvement basis. The energy sector has improved its relative productivity; notwithstanding this, the value of the EPI has remained relatively low. Thereby, the development of sustainable energy production and consumption still needs to be intensified in Lithuania. Changes in the structure of energy production sources can also contribute to the improvement of environmental performance in Lithuania.

A summary of the key findings of the research performed by the authors in terms of climate change mitigation policy is presented in Table 3.5.

The majority of large companies currently see managing their greenhouse gas emissions and reducing energy consumption as an integral part of their environmental management practices. Many have set targets to reduce their greenhouse gas emissions, in cooperation with their suppliers

TABLE 3.5

Key Findings of the Research Performed by the Authors in Terms of Climate Change Mitigation Policy

Author and Years	Key Findings
Rogel et al. (2016)	Greenhouse gas reduction measures have many socio-economic benefits. The involvement of businesses, citizens, and religious organisations illustrates the increased awareness and higher acceleration in terms of the climate.
Dimitrov (2016)	In political terms, generally, the Paris Agreement favours the developed Nordic countries that have won most of the main battles. The Agreement is less fair for the African groups and other least developed countries.
Štreimikienė (2010)	The priorities and objectives of European climate and energy policy are closely interrelated. Measures to implement climate change mitigation policy: renewable energy sources and energy efficiency.
Kelly (2016)	Climate change mitigation policy is now receiving an increasing response and public engagement.
Bassi et al. (2013)	Climate change policy is relatively relevant in today's business model and development opportunities, not only in the United Kingdom but also globally.
Baležentis et al. (2016)	The development of sustainable energy production and consumption in Lithuania still needs to be intensified. Changes in the structure of energy production sources can also contribute to the improvement of environmental performance in Lithuania.

Source: Created by authors.

and customers aiming to reduce emissions, and encouraged governments to adopt policies aimed at reducing global greenhouse gas emissions. Similarly, companies are contributing and will contribute to reducing greenhouse gas emissions through capital allocation, innovation, and the development of new technologies.

4

Climate Change Mitigation

4.1 Climate Change Mitigation Concept

Climate change mitigation involves reducing greenhouse gas emissions that are warming our planet. Many strategies for climate change mitigation can be treated as public goods: the modernisation of buildings to make them more energy efficient; use of renewable energy sources, such as solar, wind, and small hydroelectric power plants, to generate the energy; helping cities to develop green transport and promoting more sustainable use of land and forests and many others (Hardin 1968). To meet their needs, about 1.4 billion people worldwide use fossil-fuel-based energy. By 2035, the global energy demand in developing countries will increase by more than 50%. All these new consumers will need clean energy that does not harm them or the environment. The effect of climate change is visible in both natural and economic systems. The interaction between climate change and biodiversity, land degradation, loss of the wooded areas, chemicals, waste, and polluted international waters indicate the importance of recognising the consequences of climate change in all areas.

The IPCC (2007) report identifies five assessment approaches:

- impact assessment,
- adaptation assessment,
- vulnerability assessment,
- integrated assessment, and
- risk management assessment.

Adaptation assessment summarised various approaches identified by the IPCC (2007), including

- scenario-based approaches where climate risks are qualitatively or quantitatively defined and adaptation options are identified;
- technological assessments covering future adaptation options (different from those currently available) under alternative socio-economic scenarios;

- normative assessments of policy instruments using the results of vulnerability and/or risk assessments to assess acceptable adaptation options or strategies;
- risk management methods combining current risks related to climate variability and extremes with expected future developments using alternative decision-making tools to assess adaptation;
- anthropological and sociological approaches that identify the training of individuals and organisations and the processes needed to adapt effectively to the risks of climate change;
- adaptive capacity assessments, according to which investments in adaptation capacities are made in a similar way to adaptation options;
- *cost–benefit analysis*, where the benefits and costs of adaptation are expressed in monetary terms and the net benefits or costs are calculated;
- non-formalised cost–benefit analysis, which compares costs and benefits in monetary and non-monetary terms as part of multidimensional analysis;
- *cost-effectiveness analysis (CEA)* is often used to assess alternative adaptation options or the most cost-effective way to meet a specific objective (e.g., predefined ratios);
- *multi-criteria analysis* makes it possible to assess quantitative and qualitative data using several indicators;
- portfolio theory, which uses the principles of building a financial investment portfolio to maximise the expected size of the portfolio as a separate rate of return; and
- participation methods, on which the analysis of direct participation methods is based.

Before discussing detailed methods for determining costs, it is worth discussing the detailed framework that can be used in climate change analysis. An important challenge is to define a coherent adaptation strategy and prepare well-designed and flexible adaptation measures that meet various criteria, for example, efficiency (in terms of space and time), sustainability (e.g. by integrating cross-sectoral aspects), innovation (taking into account the potential of new technologies and new opportunities for climate change), and efficiency. Especially, when in the analysis the developing countries are involved, equity will be another important criterion that is also relevant in industrialised countries.

The methodology for determining the costs of adaptation measures raises a number of specific problems, such as addressing long-term risks, discounting, economic impact, and uncertainties.

There are also other related issues, including

- challenges in creating and defining the future situation (including socio-economic development);
- attributions of costs and benefits of adaptation, i.e. separation and attribution of future climate change;
- spatial and division level;
- links with conflicts of any kind;
- consideration of additional impacts or benefits of actions; and
- challenges in assessing the benefits and even costs of many non-technical ("soft") or non-market applications.

Scenario-based method of impact assessment, according to which the adaptation is assessed, requires identifying and measuring (quantitatively) the impact of natural units; converting these physical impacts into monetary values; assessing the resource costs of the adaptation options, weigh costs, and benefits of the adaptation options; and selecting the preferred option, regarding risk and uncertainty (Metroeconomica 2004).

This method is widely used. For example, this can be seen in a similar research programme ("Knowledge of Climate Change") in the Netherlands. It aims to improve adaptation challenges for various European sectors and areas, to develop, improve, and apply tools for creating coherent adaptation and implementation strategies for practical adaptation measures, and to advance the process of assessment. Tools for contributing to optimise the options are determined, and the best ranges of adaptation options are chosen.

In this way, the following scientific research issues can be addressed:

1. How to set adaptation goals in different sectors and regions of the European Union in a dynamic context, taking into account uncertainties of climate change and what would be the desired time limits?
2. How to create a coherent adaptation strategy and alternative and innovative adaptation options for both changes in the infrastructure, and behaviour and social change approach, in order to increase the European Union's resilience to climate change?
3. How assessment measures can be developed and applied to adapt across EU regions, sectors, and hotspots, including cross-cutting issues?

To consider the first issue of new socio-economic approaches, it is needed to address the questions of uncertainty, the impact of climate change, and other issues such as the optimal timing of the various adaptation goals (how much should be done and when). The second issue obliges to cooperate closely with stakeholders in various areas, to identify alternative options, ranging from technical, behavioural, and institutional options to perspectives and how this can be integrated into a coherent adaptation strategy. The third research issue focuses on urgent and remaining issues related to the costs and benefits of options, in particular optimal timing of implementation, various issues related to separate sectors, discounting and flexibility at various stages, implementation of the strategy, and appropriate adaptation options.

CEA takes a predetermined objective (often a result negotiated by the main stakeholder groups in society) and seeks to find ways to achieve this as cheaply as possible (Ahmad et al. 2001). The purpose of this analysis is to find the cheapest option for achieving physical objectives.

The classic application of CEA is to construct cost curves in order to explore the least costly ways to achieve predetermined objectives. This can be done by adapting sectoral assessments or by assessing individual risks. However, cost-effectiveness cannot be used to compare the adaptation of sectors as it was adapted to mitigation, as there are no common indicators for this. A new challenge is a recognition that climate adaptation of all human activities is very expensive, and there will be many cases where the costs will certainly outweigh the benefits. Another extreme situation is not to use any mitigation measures and to live with the risks of climate change.

The optimal policy would be the situation between these two extremes (i.e. "cost-effectiveness and proportionality"). The concept of cost-effective and proportionate adaptation is reliable; however, its practical assessment is clearly complex. Although much attention has been paid to adaptation to the effectiveness of the mitigation of climate change vulnerabilities and potential impact, it is rarely seen that if climate change mitigation measures and their implementation are selected incorrectly, the adaptation measures may actually increase the impacts of climate change. When analysing the costs and benefits of adaptation, "inadequate adaptation" shall also be taken into account as an ineffective adaptation measure.

There are several possible obstacles to identifying adaptation options in terms of cost-effectiveness:

- adaptation responds to local impact, and the benefits of adaptation are primarily local and/or regional, which is highly dependent on local conditions; and
- adaptation is a sector-specific approach to various climate signals and impacts.

There is no global or consistent data on a certain level of adaptation, i.e. it varies depending on whether the options respond to changes in average temperature, sea-level rise, or other extreme events such as floods. Therefore, it is not easy to compare the cost-effectiveness of different sectoral adaptation options or different types of measures.

Adaptation to climate change also has a cost, which is defined as planning, preparation, and implementation costs, including costs of the transition period. The benefits of adaptation can be seen as the costs of avoided damage or the benefits gained after the adoption and implementation of adaptation measures. Simply, if the benefits of economic adaptation, for example, the reduction (or potentially positive effects) of climate change impact, outweigh the costs, then there are net benefits; if not, it can lead to inadequate adaptation. This general principle is important as the resources need to be distributed effectively among different climate change mitigation strategies. This can only be done if the costs of different options are clearly identified and the benefits are assessed (note that the benefits do not necessarily have to be monetary). The cost–benefit analysis is one of the measures leading to the decision to accept or reject the choice.

Another problem with CEA is the process of selecting effective measures. A measure of effectiveness could be based on specific random individuals sample study. In practice, this method most often uses efficiency indicators selected by experts. Experts are chosen for the following reasons:

- It is assumed that experts are better-informed individuals, especially on specific issues such as habitat conservation, heritage, landscape, and others.
- Indicators selected by experts ensure a faster and cheaper change in people' attitudes (OECD 2018).

4.2 Economic Climate Change Mitigation Approaches

4.2.1 Economic Instruments of Environmental Protection

The aim of environmental policy is to strike a balance between pollution and the cost of reducing pollution. Ideally, pollution should be reduced to a level where the benefits to society of further reduction are less than the cost of pollution control equipment or reduction. Therefore, from an economic point of view, pollution should be controlled to a level where the marginal costs of further pollution reduction measures outweigh the benefits of reducing emissions. In this way, the company can balance the function of marginal pollution and the marginal cost of damage, and the state can achieve the optimum level of pollution control. In a broad sense, economic instruments are treated as "goods" in the political market (Keohane 2003 and measures of transmitting welfare.

Economic instruments are used in environmental policy for many reasons. They are based on a market price mechanism which internalises costs and creates a financial incentive for economic entities to avoid polluting. In environmental protection, economic instruments can promote economic efficiency by influencing relative cost which creates a financial incentive to change behaviour. They provide measures to internalise the costs of pollution and environmental services directly into the cost of goods, services, and activities. Such internalisation of social costs affects the price structure and thus creates an economic incentive for consumers and producers. In addition, environmental taxes are a great opportunity to manage pollution from diffuse sources, where implementation and enforcement of regulatory measures can be much more difficult. Unlike administrative controls, the use of economic instruments gives polluters an incentive to reduce pollution beyond what is required by environmental authorities.

Environmental policy instruments have been evolving purposefully. Initially, the focus was on "command and control" mechanisms, followed by a period of popularity of economic and market instruments, and now more liberal environmental measures are gaining in popularity – voluntary agreements, eco-labelling, information companies, and green purchasing.

There are various ways of classifying environmental protection instruments in the scientific literature according to the categorisation criterion. Instruments are classified according to the level of state regulation and control, resources allocated by the state, intensity of the attribute, and other criteria. One of the simplest methods is proposed by the Organisation for Economic Co-operation and Development (OECD) for classifying instruments into six types, namely:

- command and control;
- economic instruments;
- liability and damage compensation;
- education and information;
- voluntary approach; and
- management and planning.

Command and control instruments include restrictions, prohibitions, technology standards, emission limits, quotas (e.g. waste collection, packaging recycling), regulations, and certification. The regulatory objects and objectives identified by Baldwin and Cave (1999) are presented in Table 4.1.

Instruments are categorised by government-allocated resources (information, finance, coercive, and organisational) and by the intensity of the selected trait (for example, the degree of coercion). More sophisticated classification categorises environmental instruments into three categories, with corresponding symbols:

- Laws, regulations, standards – instruments that the government can use to impose sanctions and force entities to change their behaviour. Total state interference – a symbol of "stick".
- Economic measures that make governments pay entities (subsidies) or force entities to pay (taxes). Limited state interference – a symbol of "carrot".

TABLE 4.1

Objectives and Regulated Objects of Command and Control Instruments

Object and Problem of the Instrument	Objective of the Instrument and Method of Solving the Problem
External costs	Cover all expenses
Inadequacy of information	Improve the price mechanism
Public goods and market failures	Implement a cost-sharing mechanism
Rarity	Publicly based allocation
Market imperfection	Ensure efficient production
Planning	Protect the interests of future generations

Source: Created by authors based on Baldwin and Cave (1999).

- Information transfer through which governments attempt to influence operators by persuasion, knowledge transfer, and information provision (training programmes, labelling, brochures). State non-intervention, persuasion – a symbol of "instruction".

However, the most widely used difference in environmental instruments is the level of state control. When classifying environmental instruments according to the level of state regulation, it is assumed that state intervention is a key feature of environmental protection instruments. The question is whether such an assumption reflects the views of all stakeholders. It is believed that the actual level of state regulation is determined by nothing but the way of ensuring the implementation of these instruments. Therefore, the classification of environmental instruments should not be closely linked to a certain level of state regulation.

Economic instruments are the best solution for integrating external costs and correcting market failures. They are based on a price mechanism that charges external effects and provide an economic incentive for market participants to change their behaviour. In addition to taxes and emissions trading, economic instruments include public procurement, subsidies, a collateral-return system, and product tax exemptions. The elimination of energy subsidies and the liberalisation of energy prices are one of the main incentives for reducing environmental pollution and improving the quality of the environment. As the carbon market is most closely linked to taxation and emissions trading, these instruments are discussed in more detail.

4.2.2 Features of Taxes and Market Instruments

Taxes – one of the most common economic instruments. Taxes artificially increase the cost of pollution and the value of natural resources. The revenues collected from taxes are being directed to the conservation of natural resources or the mitigation of the adverse effects of pollution. Consumer taxes (e.g. waste or sewage collection) not only increase budget revenues but also influence changes in consumer behaviour. Thus, taxes have a double environmental benefit: placing a direct burden on the polluter, encouraging it not to pollute (the "polluter-pays" principle), and raising funds that are directly invested in improving the quality of the environment. On the other hand, the polluter's taxes and public expenditure on the protection of natural resources will only be matched if environmental damage is carefully calculated and preventive and remedial measures are justified. In addition, the introduction of any tax will increase administrative costs in the public sector. Opschoor and Vos investigated 14 OECD countries and concluded that target taxes account for about 0.1% of GDP. This amount is efficiently too small to turn economic development towards energy and resource efficiency. Where the burden can be justified, the "gingerbread and stick" (encourage and control) approach should be used and taxes should be purposefully allocated (von Weizsäcker and Jesinghaus 1992). Taxes are financial instruments used to distribute welfare in society. In this way, the optimal tax package reduces the tax effect, which is manifested in the distortion of price and the actions of market participants.

Economic theory states that the optimal production volume of the product is achieved when the marginal utility of consumption equals the marginal social cost. Social costs are costs that are not included in private production costs, such as damage, environmental air pollution (increased health care costs, loss in yields costs), and wage costs. In this way, the government can use the pollution tax to balance social benefits and costs by introducing a tax that is equal to the marginal or external cost of pollution. It is very difficult to determine the most effective tax level based on the marginal cost of pollution. Therefore, lately, a lot of research has been done on the measurement of external costs.

The purpose of a carbon dioxide tax or an energy tax is to increase the price of different goods in order to create an incentive in the market to reduce pollution. Pollution taxes must perform an incentive, compensatory, and cumulative function. A carbon pollution tax or a carbon tax does not eliminate carbon pollution but sets the optimal level of carbon emissions for each pollution source. This approach differs from the "command and control" principle in that the cost of environmental pollution does not dictate when and how to reach this optimal level. The imposition of a carbon tax can also be regarded as an attempt to incorporate certain external costs. The main purpose of carbon dioxide is to increase the price of different goods depending on the amount of carbon or energy. In this way, a price signal appears on the market as an incentive to reduce emissions by introducing cleaner technologies, using cleaner fuels or implementing energy-saving, efficiency-enhancing measures.

Greenhouse gas emissions are largely driven by demographic change, socio-economic development, the level, and direction of technological development. Carbon dioxide (CO_2) is the most important and most studied gas that causes a greenhouse effect. The main source of these emissions is the combustion of fossil fuels in the power generation and transport sectors.

Another relevant economic mechanism is a green budget reform. The term "green budget reform" was introduced in the 1980s. The essence of green budget reform is the introduction of an environmental tax by reducing the amount of income tax or other taxes. This results in a neutral fiscal effect and "double dividends". The aim of the implementation of this reform is that environmental taxes not only contribute to improving the quality of the environment but also generate public revenue. Due to the new revenue to the state, other taxes will be reduced while the total amount of revenue and expenditure will remain unchanged. Other taxes are treated as market-distorting taxes, and reducing taxes is seen as a means of increasing market efficiency and a second benefit of introducing pollution taxes. Thus, the impact of green budget reform can be twofold: on the one hand, it improves the environmental performance of the tax system and, on the other hand, opens up a wider perspective for improving the environmental protection tax system. Most of the milestones in terms of "green budget reform" have been achieved in the Scandinavian countries and Finland, where carbon dioxide taxes have increased public revenue. Taxes have been imposed in Finland and Norway, which have raised public revenues by about USD 580 million in each country. Currently, environmental taxes in Scandinavia and the Netherlands represent between 1.4% and 3% of gross domestic product.

It is also worth noting that the tax optimisation theory has not fully explored environmental taxes. Environmental taxes increase revenues and can, therefore, be considered as a revenue optimising measure. If a country has an optimal set of taxes in terms of revenue, there should be no change in taxes that generate the same revenue at lower market costs.

In short, a market instrument can be described as control based on a price signal. It is suggested that market instruments should be viewed as rules that influence market behaviour rather than detailed directives. The concept of market instruments has evolved as a contradiction of command and control regulations in the United States and other countries traditionally dominated by environmental regulatory mechanisms. Market forces have begun to be interpreted not only as part of the problem but also as a possible solution to the problem. Environmental protection policy instruments based on the market can be treated as a tool to overcome market failures. This definition also applies to economic instruments, which are described as instruments that make the performance of certain actions cheaper or more expensive in terms of money, time, effort, or other values. A classic example of one of the first market-oriented instruments is the Emissions Trading Scheme under the US "Clean Air Act" in 1989.

The OECD has identified the advantages and disadvantages of market mechanisms. Among the advantages the following aspects of market instruments worth to be noted:

- Under administrative control, the same measures will apply to all polluters; consequently, some polluters will have to invest more to meet the requirements.
- Market mechanisms promote "dynamic efficiency" and innovation.
- An environmental protection policy based on market mechanisms is less vulnerable to "legal traps" than policy based on negotiations at the company level.
- Market mechanisms are measures for increasing revenue.

In terms of shortcomings of market mechanisms, it should be noted that in order for environmental taxes to reduce pollution levels, polluters should respond to this financial incentive. Whether a company responds to a financial incentive or not will depend on both the company structure (if the engineering and technology structure is closely linked to financial decisions, the company will be inclined to respond to financial incentives) and the company's goals (state-owned enterprises will not always see incentives to reduce costs). The success of market instruments also depends on market models. If the market is dominated by a monopolist who seeks to maintain higher prices and profits by reducing production, an environmental tax, unlike environmental law, would cause an even greater loss of production, which would lead further away from achieving social optimum. In this case, the environmental tax would have a double effect: the benefits of reduced pollution and the loss in welfare due to the reduced production.

The "polluter-pays" principle, which states that the polluter must pay for environmental damage, plays an important role in an environmental protection policy based on market instruments. If the cost of avoiding damage is lower than the cost of damage compensation, the polluter will avoid damaging the environment. In many cases, this principle is difficult to implement as the damage that is done cannot be compensated (e.g. climate change) or legal measures can no longer be applied to the polluters (e.g. if polluter companies no longer exist) and that is why external costs arise. Quantitative indication of external costs is a difficult task. The damage can be measured by a decrease in quality of life, which is calculated as the willingness to pay for maintaining or improving the quality of the environment. In a market economy, a Pigouvian tax, covering the external costs, may correspond to green taxes, which can reach 5%–10% of GDP. However, such taxes would be a heavy burden on business and would increase social costs. The industry's response to increasing green taxes will be investments in technological innovation based on increases in raw materials' and energy prices in the long term. Therefore, environmental economic instruments (such as green tax reform) should be introduced gradually, encouraging business for long-term technological changes. The external costs of energy production can be integrated into the price of energy, through the use of pollution taxes, emissions trading, and energy taxation, or integrated into an advanced cost–benefit analysis.

The question of what determines the choice of environmental instruments and whether it is free or limited by constraints and uncertainties is extensively studied in the scientific literature. In practice, the most important criteria for the selection of economic instruments for environmental protection are economic and environmental efficiency.

Policy evaluation literature analyses environmental policies and values and identifies general policy criteria against which potential environmental instruments can be compared. The normative rationalist model is inappropriate as instruments are chosen without systematic and comparative analysis. Keohane (1999), in assessing the economic approach to the choice of instruments, highlighted a lack of empirical validity and political understanding of procedures, and presented four key criteria that are often difficult to reconcile and that determine "good governance of politics (product) and administration (process) in a democratic society":

1. Environmental efficiency
2. Economic efficiency
3. Legality
4. Democracy

Having analysed the choice of environmental instruments by comparing their environmental and economic efficiency, it was recognised that in order to evaluate the choice of an environmental instrument, it was necessary to assess not only efficiency but also the smoothness of distribution, the possibility of calculating the size, legitimacy, administrative, and technical capacity of an economic instrument (Adger 2003). Howlett (2009) analysed and compared the typologies of environmental instruments and their functions and stated that the classification of instruments is based on available financial, organisational, and human resources. Streimikiene (2005) evaluated the effectiveness of environmental tax and emission trading scheme, the cost of monitoring, the price sensitivity of entities, the readiness of entities to enter into market transactions, the possibility of implementing tax reform, the distribution effect, and the cost of welfare (where welfare costs and environmental performance were treated as the most significant factors).

It is believed that the choice of instruments is determined by both macro- and micro-economic factors. Some scientists emphasise micro-level factors: individual understanding, subjective values, cognitive factors that influence decision-makers. Table 4.2 presents the factors that influence the choice of environmental policy instruments.

TABLE 4.2

Exogenous and Endogenous Factors in Instrument Selection

Exogenous Factors			
Context as a Structure		**Context as a Problem**	**Endogenous Factors**
Systemic variables (macro)	Organisational variables (meso)	Problem-specific variables (meso)	Individual variables (micro)
Characteristics of national policies Culture of politics Social fragmentation	Organisational culture (method of employment, life expectancy, method of socialisation) External contacts (clientele, political communities)	Policy (volume) Requirements (symbolic, functional, support) Restrictions (commitments, resource constraints)	Roles (professional, institutional) Cognitive factors (ideology, values and criteria, perception)

Source: Created by authors.

The valuation of economic instruments presents several challenges. When analysing the impact of an instrument or policy instrument, the benefits and costs of possible instruments are assessed. The effectiveness of an environmental instrument will be assessed by comparing the situation before and after the instrument has been implemented with respect to its environmental impact. On the other hand, the costs of the introduction of the instrument and its possible side effects need to be considered. Therefore, when evaluating different environmental instruments, it is essential to evaluate the costs and side effects. By analysing the effectiveness of instruments, environmental performance, cost-effectiveness, and administrative and compliance costs, revenue generation and other economic impacts (price level, competitiveness, trade structure, labour market, income distribution, economic growth, and the level of innovation) are assessed.

Although monetary value is a decisive criterion in the choice of instruments for reduced environmental damage, the uncertainty in the assessment of this criterion is greater. It is easier to obtain quantitative information on emissions than to assess the impact of a measure on reducing environmental damage. Ex ante (before the introduction of the instrument) and ex post (after the introduction of the instrument) the assessment may vary significantly due to the difference in the amount of available information. This happened with the Emissions Trading Scheme (ETS), where a much lower price of ETS was expected before the system was introduced. In addition to the direct environmental impact of the measure, side effects should also be considered (e.g. reducing sulphur dioxide will also reduce greenhouse gas emissions). On the other hand, future benefits should be discounted, though it is difficult to estimate future benefits in terms of initial costs or current benefits. Furthermore, if several instruments are used to address an environmental problem, it is difficult to distinguish the environmental impact of each. The effectiveness of the instruments will be influenced by whether or not polluting companies are allowed to fail to meet environmental standards. If the company has the opportunity to go against environmental requirements, the way to minimise the company's costs will change – companies will be able to emit more pollutants than allowed.

In assessing the environmental effectiveness of market instruments, it is especially important to assess the benefits and costs of their introduction and implementation as environmental effectiveness depends on the response of polluters to the market signal. By applying administrative instruments and legalising quantitative restrictions on pollutants, the environmental impact depends to a large extent on the implementation of the measure, yet is still more reliable compared with the environmental impact of economic instruments.

There are three ways to evaluate environmental effectiveness:

- reduction of pollutants measured by physical parameters (e.g. waste t/year);
- the reduction of environmental damage due to reduced pollutants measured by physical parameters (reduction of pollutant concentrations in adjacent water bodies or in the atmosphere, e.g. CO_2/year); and
- the economic value of reduced environmental damage in monetary terms.

The advantage of economic instruments over command and control measures is the lower cost of achieving the same environmental result. Therefore, when evaluating the economic effectiveness of instruments, it is important to calculate whether the instrument minimised the cost of reducing pollution or not. These are the direct economic costs resulting from a change in the polluter's behaviour. These costs include the cost of pollution abatement equipment and operating costs of new technologies in industrial enterprises. The effect of an economic instrument on a household is measured in terms of household expenditure and costs of changing the structure of consumption (towards the least polluting products or services). If there is a significant difference in the cost of reducing emissions between different polluters, market participants with lower pollution abatement costs can reduce more by removing the burden on other market participants. This is particularly relevant to the expansion of the EU Emissions Trading Scheme when it comes to integrating the aviation sector, whose cost of pollution abatement is significantly higher in comparison with other sectors.

Scientists are extensively analysing what kind of impact on competitive business has environmental requirements. In assessing the impact of environmental requirements on the competitiveness of enterprises at the international level, foreign direct investment is the most commonly analysed. Enterprises are resisting and delaying the implementation of complex (extensive) environmental requirements as they increase fixed costs. The negative impact of environmental requirements on trade flows and competitiveness results in polluting industries being moved to countries with less stringent environmental requirements. Scientists often base such findings on the Heckscher-Ohlin model of international trade, which ignores the dynamic outlook, future cost uncertainty, technological development, and strategic behaviour of enterprises. The traditional approach is based on the idea that a compromise must be found between the social benefits and the businesses' private costs in reducing pollution, introducing preventive measures, and cleaning the environment. These costs will be compensated by raising the price of an enterprise's products and thus reducing its competitiveness. Future foreign direct investment is also diminished by the uncertainty of future environmental requirements, and the most important environmental factor determining foreign direct investment is the predictability of environmental requirements. Opportunities for energy-intensive businesses to become less competitive, relocation of businesses from the European Union due to the introduction of EU ETS, and competitiveness of the EU in global markets are widely discussed by scientists regarding the effectiveness of the EU ETS.

Having considered the impact of environmental policy instruments based on the market structure, it can be concluded that it is inefficient to use a single instrument in environmental policy and it is necessary to adapt the instruments to the specifics of the market and to select a number of instruments to cover the side effects of international relations between market participants (businesses, governments).

4.3 Emissions Trading Theory

The Coase theorem stating that market participants can reap the benefits of trading without government intervention is valid only if the property rights are well defined and if there are few participants in the market. In reality, there are not many such cases and, therefore, when property rights are difficult to define and the market is affected by a large number of participants (polluters and the aggrieved), an economic instrument that is market-friendly is needed. In such cases, the Pigouvian tax could be applied or an artificial market for pollution allowances could be created.

Trading of emissions is based on market forces and is suitable for reducing aggregate pollution in a large area. Therefore, such an economic model of the instrument is well suited to addressing global environmental challenges. Montgomery theoretically proved that the system, when exposed to market conditions, allows achieving environmental goals at the lowest possible cost. There are theoretical and practical examples of how the markets of sulphur oxides and nitrogen oxides work to combat acid rain. While addressing climate change, a carbon emissions trading market is being created. Representatives of the industry believe that trading of emissions is more flexible than legal regulation and a future emissions trading will play an important role in corporate choices and encourage businesses to choose the lowest-cost emissions reduction method and promote green technology development in developing countries.

Trade and environmental policy are closely linked. With regard to trade theory and environmental economics, researchers discuss the impact of national environmental policies on the competitiveness of enterprises on a global scale. Governments tend to establish lax environmental policy, while others, on the contrary, argue that environmental policy will be strict and will thus promote innovation. Both positions relate to the possibility that governments will deform their environmental policies for strategic reasons. To verify these claims, the modelling of environmental policy in terms of imperfect competition is needed, where the government achieves strategic goals through market manipulation and environmental policy, while manufacturers manipulate the market by choosing a geographic location and by using researches and development. According to Krugman (1991), these factors are closely related to the theory of international trade, the theory of multinational corporations, and the theory of geographical distribution of economic activity. These theories are used by environmental economists to analyse the impact of environmental policy on the open economy.

Scientists have developed a number of theoretical models of emission trading systems operating assumptions that the market for allowances is perfect and therefore the costs of reducing emissions will be the lowest, or considering trade as the best option. The main focus of all models of international emissions trading is placed on the efficiency of the system and the influence of the methods of allocation of allowances.

There are several alternative forms of the market for emissions' allowances, which have been analysed in detail in Čiegis (2004). Environmental-permit system forms a portfolio of allowances, while permits are issued not on the basis of emission sources but regarding the impact of these emissions on a given geographical area. In this way, the number of emissions that cause more damage at any particular point must be purchased/sold to compensate for more emissions from another source, the emissions of which per unit account for a lower proportion of the pollutant concentration at those points.

The emissions-trading system is based on a single emissions limit in a region and provides for a single market for eco-licences and a single price for emission allowances. In this case, the region will be divided into zones, and within each zone, the sources of pollution will trade in emission allowances within the region. The emissions-trading system simplifies the exchange between sources and reduces the costs of finding partners, servicing transactions, and forecasting the prices of eco-licences. On the other hand, the emissions-trading system cannot ensure that the lowest cost solution is chosen.

The third form of the emission allowance market is a mixed system, a system of counterbalancing emissions, which provides for a limit on emissions per time unit and sets boundaries of the area where trading is possible. A system of counterbalancing emissions is the most preferred model for the development of emission allowances systems. In such a system, new sources of pollution should compensate for emissions by reducing them in other sources (e.g. terminate the operation of polluting installations, invest in energy-saving or public transport or reforestation). Such a system for reducing carbon dioxide emissions encourages investment in energy efficiency measures and energy savings and the search for cheaper ways to reduce carbon emissions. The system of counterbalancing emissions is an appropriate instrument for large point pollution sources, however, not appropriate for the control of pollution from small and scattered point sources (road transport, households) due to high administrative costs. It is proposed to apply a carbon tax to these sources of pollution and to use the revenue collected to finance the counterbalancing system.

In the market for the trade of allowances, the following trading system models are available: "cap-limit", "compensation", "network", "bubble", and savings.

In the "cap-limit" system, the state sets a permissible emissions limit for a polluting object, above which the polluter must pay a fine, or has the possibility to purchase the right to pollute from other commercial objects. The "cap-limit" model was first implemented in the US "Acid Rain" programme in the 1980s. The "cap-limit" system has been found to have economic and environmental benefits. The impact on the quality of the environment has been assessed qualitatively, and although in many cases it is not significant, there has been a significant improvement in the implementation of the legislation. On the other hand, one of the main factors limiting emissions trading in the US trading system was the transactional costs (administration fees, documentation, finding a trading partner). The "cap-limit" model is based on both the EU ETS and other emerging greenhouse gas trading systems.

Compensation model is applied in the region where a new business-polluter wants to start its activity; however, the overall level of pollution in the region cannot rise. Under the compensatory policy, a start-up in a region must "compensate" additional emissions by purchasing emission reduction credits from a business already in the territory, which automatically reduces its emissions by the same amount.

Network model is used when the reconstruction–development of a business is being carried out. Trading shall then take place within one consortium, where emissions from one source of pollution are transferred internally to another source within the consortium, but the total emissions from the consortium remain unchanged.

As regards the bubble method, the sources of pollution belonging to a single object or located in a single area are aggregated and controlled not by the emissions of each individual object, but by the total emissions. "Bubble" emissions allow businesses to increase their flexibility to meet environmental standards and allow businesses to save.

Savings model is based on allowing a business to save allowances and sell them to another company in the future. Such an environmental policy encourages the business to reduce emissions beyond what environmental standards require in order to be able to emit more pollutants in the future (due to increased production) or to generate additional revenue.

Trading of emissions is a cost-efficient instrument. Thus, the buyer of the emission allowance wins by saving costs, and the seller wins by obtaining a profit on the sold emissions allowance. The marginal cost of reducing pollution in a company will be a decisive factor in choosing the role of buyer or seller in the market.

The impact of the ETS on the market (businesses and corporate customers) will depend on a large extent on the market for each business's product and how the change in cost will affect the price and quantity of the product. This change will determine the following:

- The price of allowances will determine the costs of compliance with the limit, and the price itself will depend on the total amount of allowances (surplus or shortage in the market).
- Alternative costs – whether allowances can include additional costs resulting from the price of allowances in the price of the product.
- Market conditions – whether the business operates on the international market (where allowances will not affect prices) or on the national or global market (where the price of allowances will affect the costs of businesses).

Costs minimisation in perfect and imperfect market conditions will be different. In conditions of perfect competition, where no market participant has the power to influence the price of the emission allowance, all market players will purchase the allowances at the same price as their marginal disposal costs. The costs of disposal will then be the same for all market players, and the overall costs of disposal will decrease, regardless of the initial allocation of allowances.

In terms of imperfect competition, a single business can influence the price of an allowance and reduce its costs. In other words, it will use costs reduction manipulation. It is important to determine how much the system decreases efficiency in terms of the costs reduction manipulation and found that the influence on effectiveness is based on three factors: the allocation of the initial allowances; the dominance of businesses that cannot influence the price in the market; and the specifics of costs function. On the other hand, in terms of imperfect competition, the dominant business may seek to reduce the competitiveness of other existing or new businesses (so-called "exclusion manipulation").

In assessing the efficiency in the trading of allowances, it is proposed to analyse the turnover of allowances, the price of allowances, the geographical distribution of emissions, and the admission of new entrants into the market, the introduction procedures, and their costs. If there are few buyers and sellers in the market, an issue of the competitiveness of the market appears. Low turnover of allowances may indicate that trade regulations are too complex (e.g. complex trade regulations in the United States have increased transactional costs and reduced trade volumes). In terms of low allowances turnover in the market, a significant fall in the costs of pollution reduction will not be achieved, as the price of allowances should reflect the costs of pollution reduction. On the other hand, the higher the price dispersion, the lower the efficiency of the market is.

In assessing the effect of geographical emissions redistribution, it is worth noting that if the location of the pollution reduction is not affected, the geographical dispersion of the pollution reduction will increase the efficiency of the system (e.g. in case of greenhouse gases). If allowances are allocated free of charge to existing market participants and new entrants must buy allowances from the existing market participants, the new entrants will not have an incentive to enter the market due to higher costs and adverse conduct by the existing businesses to its competitors. In any event, it is very difficult to estimate how many new businesses would have entered the market without the introduction of the ETS. The assessment of the introduction costs shall take into account the monitoring of implementation, the costs of introduction, and the level of the compliance with environmental requirements. In addition to the above-mentioned criteria, the assessment of economic instruments shall take into account the impact on change in attitudes and the introduction of new, more advanced technologies.

In theory, emissions trading has a number of advantages. Cost savings are achieved because all participants involved in the trade save costs and the government does not incur additional costs. Environmental benefits are achieved too; as the number of allowances is predetermined, the uncertainty that emissions will be reduced (unlike in the case of taxes) disappears. If a new installation appears on the market, no new allowances will be issued, and the price of the allowance will automatically rise. For new installations, the allowances can be bought from existing ones, thus combining economic growth and environmental quality improvement. In addition, emission allowances prices are subject to inflation, and there is no need to adjust them, unlike in the case of taxes.

The emissions trading system (ETS) has a positive impact on the dynamics of the economy. The demand for emission allowances depends on the emergence of new businesses, the closure of existing businesses, and technological progress. For businesses, trading allowances is a convenient way to adapt to changing market conditions. In a growing economy, emission allowances are a more favourable instrument than the Pigouvian tax, since the price of the emission allowance will automatically increase if the quantity of the allowances in the market relatively decreases, and the Pigouvian tax should be adjusted periodically. Emission allowances are also a more flexible instrument in the event of a declining economy. If the Pigouvian tax was not adjusted in a declining economy, the environmental quality would deteriorate, and in the case of emissions trading, the allowance price would correspond to the overall price level. At the business level, emission allowances provide the most cost-effective way of reducing emissions.

Despite the above-mentioned advantages, emissions allowance models have a number of challenges to address in climate change policy. Firstly, the lack of scientific validity for the causes of global warming will lead market participants to question the expedience of pollution reduction. The decision on targets of the pollution reduction will be conditioned by a political compromise. On the other hand, careful measurements and calculations to control actual emissions are necessary to ensure the reliability of the system. There are a number of conditions necessary to successfully apply emissions trading in practice. Firstly, the market participants' marginal disposal costs must differ. The greater the difference in costs, the greater the benefits for market participants are. However, a sufficiently large number of players must be active in the market. Polluters must have the potential (economic and technological) to reduce the costs of disposal. The success of the system is also determined due to clear allocation rules and the ability to identify and measure emissions, and the authorisation of allowances must be implemented.

Transactional costs are one of the potential weaknesses of the ETS. In the ETS, the search for a trading partner, costs of seeking and obtaining information, costs of negotiation (broker fee), legal costs, and insurance costs result in transactional costs. Transactional costs will reduce emissions trading. In addition, the overall cost of reducing the pollution will be higher compared with the lowest cost option in the absence of transactional costs. Transactional costs can be reduced by an appropriate emission trading structure. According to Tietenberg, transactional costs can be reduced by providing detailed information to market participants and making market prices public, enabling buyers to find sellers and transact between themselves. Insufficient information on prices complicates negotiations and the planning of investments in reducing pollution. Since the price of a pollution allowance should be in line with the marginal costs of control, the information on these prices and their dynamics would, therefore, be valuable for investment decisions in reducing pollution. The lack of information can be addressed by launching "local" and future auctions. The establishment of exchanges where buyers and sellers can learn about trading opportunities can also reduce transactional costs.

It is worth noting that in recent decades, the European Union's environmental policy has shown a trend towards a reduction in direct state government interference. It is therefore natural that the European Union has introduced in the climate change policy an economic instrument – emissions allowance trading. Global interest in market mechanisms for climate change emerged with the adoption of the Kyoto Protocol in 1997. Business attitudes towards climate change strategies have also changed: in the early 1990s, climate change strategies were dominated by political rather than market aspects, the significance of which subsequently increased considerably. A wide range of international options have emerged to limit corporate greenhouse gas emissions: the European Union and global emissions trading and other flexible Kyoto mechanisms.

The modern theory of organisations states that the behaviour and attitude of the organisation and the individual is determined by a set of external and internal factors. Internal factors are significantly influenced by external factors and vice versa. The organisations are open systems integrated into the social context, symmetrically or asymmetrically depending on the external factors, and affecting the environment through external uncertainties (complexity and change) and external pressure (values, culture, politics, and technologies). To endure and work profitably, organisations need to adapt not only to the changing physical environment but also to the institutional environment. It is the institutional environment that determines what is considered legitimate.

When formulating an environmental strategy for a business, it is necessary to analyse the external and internal factors. External threats and opportunities (impact of legislation, market pressure factors, opportunities for new markets, or cost reduction) and internal strengths and weaknesses (environmental impact of a business, energy consumption) must be analysed. Then it is possible to develop a strategy that pursues three environmental objectives: to make existing impacts acceptable, to identify and realise the existing potential, and to change activity towards sustainable business (Table 4.3).

Taking into account the carbon market, fuel market, electricity markets, a close relationship between climate change policy and regulatory uncertainties, and the diversity of national regulations on the distribution, the inclusion of an element of the carbon market to the investment decisions becomes a complex challenge for businesses.

Under different historical, economic, business, and social environment in the EU member states, the effects of the implementation of the EU ETS will be different in different member states and will differ depending on whether the state is a potential "buyer" or "seller", which determines the designated quantity of the allowances to the business, which shall be determined in the national plan for the allocation of allowances. A state's legal obligations, national and international, state's geopolitical situation, production of energy resources and energy resource structure, and the certainty of the ETS trading rules are among the external determinants of a business's choice.

The ETS market and the energy market are closely linked. Market activity will be influenced by the development of the market for energy resources

TABLE 4.3

Factors and Variables Influencing the Management of Greenhouse Gases in a Business

Factor	Variables Determining the Factor
Management of allowances	Market price of allowances Potential reduced GHG amounts in the business Price forecast Activity in allowances trading Emitted GHG reduction scenarios within the business Market access opportunities Taxes
Compliance with the requirements	Requirements for environmental allowances Monitoring of the amounts of emissions Data quality assurance Preparation and presentation of reports Assessment of greenhouse gas reports Account management in the register
Other aspects	Legal harmonisation GHG budget/imposed strict-reduction allowances Knowledge of the business

Source: Created by authors.

(gas, carbon, oil), demand for energy, and interconnection of energy networks. In addition, the ETS market is affected by weather conditions. Drought reduces hydropower production and increases fossil fuel burning and emissions accordingly. In assessing the overall impact of weather conditions, into account must be taken not only rainfall but also temperature. In addition, it should be noted that conditions in all the EU hydropower regions (the Iberian Peninsula, the Alps, and Scandinavia) need to be assessed since rainfall per country will differ significantly from the overall impact on the European Union as a whole. The model of external (exogenous) and internal (endogenous) factors influencing the carbon market is shown in Figure 4.1. Exogenous factors are closely linked to the energy market and internal factors – to the ETS market.

The strategy of the businesses will also be determined by the rules and development of the ETS itself. When making investment decisions in a business, it is necessary to predict the development of the ETS, the impact on costs of the changes needed to meet environmental requirements. Depending on whether the amount of ETS allocated will remain unchanged in the future or decrease in a few years, the company may make alternative investment decisions. In addition, business planning is complicated by the short-term nature of trading periods (3 years and 5 years), and lending and borrowing insurance.

Emissions trading will directly affect the business's strategic positioning and asset management, carbon intensity, the geographical location of the business (domestic fuel balance, cross-border electricity connectivity),

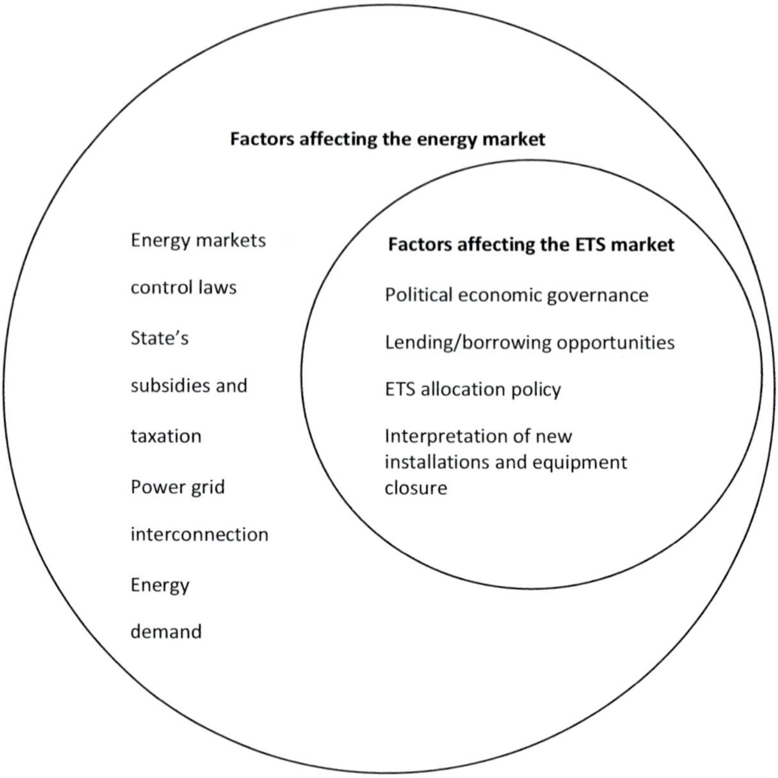

FIGURE 4.1
Factors affecting the energy market and the ETS market. ETS, Emissions Trading Scheme.

Source: Created by the authors according to PricewaterhouseCoopers 2004.

financial incentives to use less polluting fuels, and a ratio of emission reduction costs to the price of allowances.

The business's position on the market will be determined not only by the number of allowances allocated but also by the business's greenhouse gas emissions. It is important for the company to know production volume forecasts and carbon dioxide emission projections as well as the relationship with the allocation of the allowances. Considering this, it will be planned whether a surplus of allowances should be sold on the market or the surplus of allowances should be purchased at the lowest cost. Short-term and long-term strategies differ only in the level of uncertainty over time: the level of uncertainty increases. The better the company will know the GHG emission quantities of the forecasts and level of uncertainty (due to fuel prices, carbon prices, capacity substitution), the more the company will be able to represent itself, create marketing strategy on the ETS, and take advantage of the fluctuations in the price of the benefits. In theory, adding the carbon price to the dynamic projection model for the GHG emissions can be considered as emission quantity projections. In practice, investment decisions will be based on a system of criteria that is difficult to define by simple calculations used to predict the GHG emissions.

The techno conceptual options for reducing greenhouse gases in an undertaking are defined by internal investments in energy efficiency and fuel replacement. The most commonly used are such techno conceptual methods of reducing direct emissions:

- change of fuel type used to less polluting (from heavy fuel oil to gas or biofuels);
- development of renewable energy sources;
- implementation of energy efficiency measures (e. g. renovation of existing installations);
- change in processes;
- replacement of products; and
- capture and removal of carbon dioxide.

The industry should reduce emissions by introducing new technologies, based on the best available method of manufacturing. There is an economic incentive for such investments, since the allowances saved can be sold on the market and thus recoup some of the investment in new technologies. First, the easiest and most effective measures are taken, and the most significant reductions are made. In the longer-term perspective, the potential for reducing pollution increases, resulting in higher emission reductions.

Businesses burning biofuels in their boiler rooms instead of fossil fuels are particularly favoured since biofuels are considered carbon-neutral fuels and all allowances allocated to the replacement plant will be sold. According to PricewaterhouseCoopers (2004), switching from carbon to gas reduces the cost of emissions by EUR 5/MWh (assuming 500 kg CO_2/MWh at a CO_2 price of EUR 10/t).

The business's response to the market is part of the business's broader strategic position, which includes both market and nonmarket components. The response will depend on the economic significance of the EU ETS for the business. If the costs of allowances represent a small part of the total costs of production (e.g. in the case of the food industry), the resources allocated to the strategic response will be correspondingly small. In such industries, it is common practice that the impact of fluctuations in electricity or fuel prices will be much more important than the direct costs of reducing GHG. Products that are produced by emitting high carbon emissions (e.g. electricity generation by burning carbon or fuel oil, the cost of carbon can be up to 25% of total production costs). In such industries, carbon management should be given greater attention.

The participation of the business in trading allowances depends on the attitude of the business's management towards the opportunities of trading allowances. The activity of the business in the market is influenced by responsible employees in the business, the level of responsible employee in the corporate management hierarchy, and his/her skills. Actually, the business must have a person in charge of the management of emission allowances, who can be either in the environmental unit or in the trading unit.

To assess of the costs of emission reduction, usually, the projections of the development scenario of an emitted amount of the GHG without additional measures are defined and the possible options for reducing the amount of the GHG are calculated. It should be assessed whether the business has installations outside the ETS (e.g. installations of less than 20 MW). This may include installations owned by the business or its partners that are located outside the European Union. In the case that the greenhouse gas emissions are reduced in the non-EU member states, then the number of allowances may be taken into account through Joint Implementation (JI) or Clean Development Mechanisms (CDMs).

The business knows the amount of allowances to be allocated in advance; nevertheless, other factors are variable. For example, the production of energy undertakings, as well as pollution, depends on climatic conditions, and industrial undertakings are dependent on market conditions; therefore, it is quite challenging to predict the development of some industrial undertakings. At the same time, the market price of allowances is volatile and uncertain (it is clear that the market price of allowances will not exceed the fine of EUR 40/t CO_2 for the shortage of allowances). Moreover, emissions trading will not take place at the national level, but at the European level, which further increases market uncertainty. Having foreseen future allowances prices, an assessment of the actual carbon emissions and the allocation of allowances to the undertaking, and the potential for carbon reduction, the position of the undertaking on the carbon market might be formed, and business operations and investment strategy might be adjusted by including buying and selling. Taking into account the carbon market, fuel market, electricity markets, a close relationship between climate change policy and regulatory contingencies, and the diversity of national rules on the distribution, the inclusion of an element of the carbon market to the investment decision becomes a complex balancing challenge for businesses.

In summary, it can be said that emission allowances trading gives the company the flexibility to create the same, the most effective strategy for estimating the number of factors: the business approach to trade emission allowances, as a new capitalisation of the area, existing and future carbon emissions, amounts of distributed allowances, technoconceptual emissions reduction options, and carbon market dynamics and prices (the fuel market prices dynamics), taxes, and potential legal changes of requirements.

In environmental management, the market response is defined as the strategic behaviour of businesses in maintaining a competitive advantage they already have, or in gaining an advantage, through environmental factors. Proactive environmental behaviour is likely to provide a competitive advantage. This approach demonstrates the internal interest of strong companies, arguing that environmental response enables the organisation to develop strategically valuable capabilities that lead to competitive advantage.

The businesses' greenhouse gas management strategy is influenced by the business's actual amount of emissions, assigned amount of allowances, internal reduction possibilities, and external reduction possibilities through the acquisition of emission allowances and emission reduction units (ERUs). Based on the prices of allowance and ERU, the business will choose a greenhouse gas management strategy that will be aligned with and integrated into the business development and investment strategies. In this way, the greenhouse gas management strategy will become an integral part of the business development strategy.

Emissions trading allows businesses to choose the ways to reduce pollution: modify products or processes; buy allowances; buy external carbon credits generated while participating in flexible Kyoto Protocol mechanisms (JI projects or CDMs). This question relates to the question of transaction theory as to why businesses choose to produce a product or service themselves rather than buy it on the market. The answer depends on the relationship with other members of the system.

The choice of the business whether to produce the product or buy it in the market is determined by three factors: the level of transaction uncertainty/complexity, the density/frequency of transactions, and the carrying out of the necessary investments for transactions (operations performance, partner finding, negotiations, lawyers, and expert consultation). If the transactions are complex and the level of uncertainty is high, then, according to Williamson, the remaining two factors will determine whether the business will choose the market opportunity and buy a product, or will choose the hierarchical solution and will produce the desired product itself. If there is little specific investment and the frequency of transactions is low, organisations may opt for a market solution, even when conditions are not defined. If specific costs are low but the frequency of transactions is high, a bilateral agreement with a clear contractor will be more attractive. If the frequency of transactions is low and the specific costs of transactions are high, partners will face the risk of strong interdependency. In this case, a third-party solution should be sought. If both the frequency of transactions and the specific costs of transactions are high, the most rational option would be to hierarchically integrate the product within the business. This would reduce the level of uncertainty and transactional costs.

In the context of emissions trading, the analysis of transactional costs will focus on

- different options for achieving the greenhouse gas levels set out in the national allocation plan;

- how often does a business will require "place"/greenhouse gas emissions allowance; and
- the volume of the specific investment necessary for transactions depending on various choices.

The market choice will be made when the business decides to buy/sell allowances in the EU ETS, and the bilateral agreement is made when the business decides to invest in JI and CDM, and to obtain credits, which will be able to use in the EU ETS. A hierarchical solution will be considered as a solution to reduce GHG emissions within the business. A third-party solution could be considered as an investment in a fund (World Bank Prototype Carbon Fund (PCF)). It is not clear how attractive a bilateral agreement model will be for businesses, especially if the business participates in the EU ETS.

According to the response to changes in environmental requirements, businesses can be divided into four categories: "ostrich", "lagging behind", "contemplating", and "acting" ones.

Depending on external political, institutional, market factors and internal structural factors in the business, the business's management body can choose several strategic positions:

1. not to comply with legal requirements (to expand production without reducing GHG and to emit more GHG than authorised in the allowance);
2. meet the requirements with a minimum contribution;
3. optimise the price of allowances (carbon reduction options, emissions forecast); and
4. receive added value by trading allowances.

The key condition for maintaining competitiveness is cost reduction and differentiation of products and services. These measures gain businesses a competitive advantage. There is limited scope for product differentiation in trading allowances, as the types of products produced are restricted by an EU directive. Therefore, the most important internal factor for competitiveness in the system will be costs reduction. Businesses can take a variety of costs reduction measures, such as changing the used raw materials, waste management, distribution and packaging, and improving energy efficiency.

4.4 Environmental Management Systems

The term "quality system" originated around the 1950s. Quality control, as an element of quality management, was formed after World War II. The basic principles were systematised by J.M. Juran in his book *Quality control handbook* in 1951. In 1959, the first quality programme requirements standard Mil Std 9858 was released in the United States. This standard formed the basis for all subsequent quality system standards.

An environmental management system (EMS) is a tool designed to continuously improve environmental performance in all types of organisations. EMS is a system for identifying and addressing environmental problems which contributes to the achievement of environmental commitments and efficiency improvement objectives. The EMS provides the opportunity to systematically manage the direct and long-term impact on the environment of organisation's products, services, and activities, as well as consistently address current environmental issues, funding, defining responsibility, and continuously assessing the activities, processes, and procedures. ISO (International Organization for Standardization) defines the EMS as "part of the overall management system, that includes organisational structure, planning activities, responsibilities, practices, procedures, processes, and resources for developing, implementing, achieving, and maintaining the environmental policy".

The EMS provides a framework to systematically respond and manage immediate and long-term organisation's products, services, and processes impact on environment and allocation of resources, attribution of responsibility, procedures, practices, and processes of continuous evaluation assistance, which allows solving environmental problems in an organised and consistent manner.

The EMS can be implemented in many ways. It may depend on the specific sector or operations, the need felt by the management body, yet in all cases, there must be some common key elements: environmental policy, environmental programme or action plan, organisational structure, integration into operations, documentation system, necessary to collect, analyse, monitor, and restore corrective and preventive actions, EMS audits, management analysis, training, and public relations.

Quality management – part of the general function of the organisation, which determines the quality policy, objectives, responsibilities, and quality management measures and has the following five elements: planning, management, assurance, control, and improvement (Ruževičius 2007). The quality management system is a set of organisational structures, responsibilities, duties, procedures, processes, and necessary resources for effective quality management of activities. This system is usually documented and integrated into organisation's technical and general management systems, in order to sustainably coordinate and reconcile all the actions leading to product and service quality, consumer satisfaction, operational cost reduction, and financial results. Quality management models and methods are also used to solve environmental and sustainable development problems. EMSs essentially are quality management systems that focus on implementing preventive environmental policies.

The principle elements of an EMS are (Maier 2005; Cascio 1999; Marcus and Willing 1997) environmental policy – should be made publicly available to the society, in order to show a commitment to address a company's environmental impacts; environmental protection realisation programme; legal requirements and environmental aspects associated with organisation's activities, products, and services settings; the environmental planning encouragement in all organisational activities – from raw material extraction to distribution of the product; identification of significant environmental impacts – a company should analyse its activities to identify those areas where it has most impact, for example, energy consumption, emissions to air, water pollution, waste, and water consumption; setting measurable objectives and targets – the policy should include a general commitment to minimise the impact on the environment; documenting processes, procedures, and responsibilities – the company should establish a liability structure within the organisation, as well as documented procedures and manuals to support the system; internal auditing – to include internal audits and site audits as appropriate; internal reporting and management analysis – the company should ensure efficient environmental control, supervision, and monitoring, as well as a continuous analysis of the EMS (see Figure 4.2).

By incorporating EMS to its management system, the organisation has the opportunity to balance and combine economic and environmental interests by creating an integrated management model (Figure 4.3), thereby gaining an advantage in competition.

EMS is based on the concept of continuous improvement, which includes the principle of measurement and evaluation. This principle states that an organisation must measure, verify, and evaluate its environmental performance.

FIGURE 4.2
Elements of the environmental management system.

Source: Created by authors based on Maier 2005, Cascio 1999, Marcus and Willing 1997.

FIGURE 4.3
The Deming cycle.

Source: Created by authors based on Moen and Norman 2009.

The EMS deployment process is influenced by Plan–Do–Check–Act standard philosophy, which is based on the Deming cycle (Figure 4.4).

The EMS begins with the development of an environmental policy, which includes the organisation's provisions. In the LST EN ISO 14001 standard, several steps of the operation of the system are identified (Figure 4.4):

First step is planning: At this step, a company must set and continuously review its activities, products, and services that affect the environment. To implement it, it is necessary to review all activities of the company, territory, used materials, technological processes, generated waste, and where it is transported to substantially. This way, aspects that have a significant impact on the environment are identified. According to existing mandatory requirements, significant environmental safety aspects, technologies and financial capabilities, operational and business factors, and parties' interests lead to environmental goals and objectives.

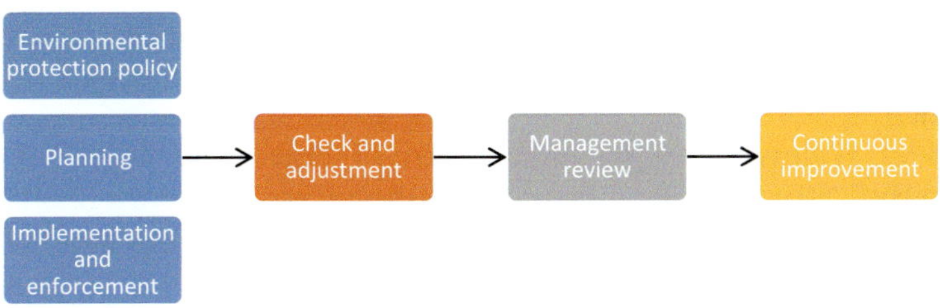

FIGURE 4.4
The key elements of ISO 14001.

Source: http://isoupdate.com/resources/five-key-elements-for-a-successful-iso-14001-management-system-ems/.

Second step considers the implementation and execution of the environmental EMS. To this end, a company establishes a structure, capable of managing environmental measures, which requires human, technical, and financial resources. Employees of the company must be trained and prepared in the field of environmental protection. Main elements of the environmental EMS are described, i.e. requirements described in the documents for efficient implementation of systems activities, its integrity, environmental policy, and objectives achievement. A company shall also evaluate the likelihood of environmental accidents and prepare itself to prevent them or to reduce their impact.

Third step is verification and correction. A company must set up monitoring and measurement procedures and observe data to ensure that they do not deviate from regulations set by the country on environmental protection.

Fourth step takes an analysis of system performance. The periodical analysis is performed by the company's management. The information, collected on the EMS operation and the compliance of environmental protection indicators, determines management decisions on the change of environmental protection policy, goals, or other EMS components in order to achieve better environmental protection management system efficiency.

EMS can be certified when it is installed and operating; generally, stages do not differ from the certification of the quality system. Internationally, the main organisation for standardisation is ISO. ISO is a global network that identifies which international standards are required for business, government, or society and has 157 member countries. The main objective of the organisation is the development of international standards to encourage international trades and technical progress.

Since the start of its activity in 1947, the international organisation for standardisation has published more than 19,000 international standards covering a wide range of fields (À propos de l'ISO. 2014). Since the beginning of the ISO 9000 series standards, it was implemented by many organisations belonging to various industrial sectors (Franceschini et al. 2006).

When developing ISO standards, ISO instructions are used, according to which the content of documents must be arranged exceptionally seamlessly. Standards are presented in a way that they can be adapted to national standards without additional conversion. Documents are published in English and French.

ISO 14001: Following the successful introduction of the Great Britain EMS standard BS 7750 in 1992, European and international standard EN/ISO 14001 was released in 1996. Since then, tens of thousands of organisations around the world have already used this standard to solve their environmental problems, including over 30 Lithuanian companies. In addition, better control of materials and energy costs and an improved image have also contributed to business development. All the requirements of ISO 14001 are included in the EMAS Regulation; therefore, ISO 14001 standard can be useful for organisations that are seeking EMAS.

EMAS: European Management and Audit Scheme is a system based on voluntary participation for companies and other organisations that wish to commit themselves to assessing, managing, and improving their environmental performance. The system was created in 1993 and updated in 2001. EMAS is similar to ISO 14001, which, however, poses higher requirements: EMAS-registered organisations must demonstrate compliance with environmental protection legislation requirements and to present information about significant environmental aspects and associated environmental aims and objectives, key data on environmental performance publicly.

Environmental Management and Audit Scheme (EMAS) enables an opportunity for organisations which are operating in the European Union and the European economic area, to participate freely in the EMS. The scheme began to operate since April 1995. It seeks to promote continuous improvement and assessment of environmental performance for participating organisations.

For a different type of organisations, different types of EMSs are used. It depends on the kind of organisation, volume and structure, production, and provided services. All EMSs have common components. The main components are the following (Staniškis and Stasiškienė 2007):

- The environmental protection policy, commonly referred to as the environmental protection policy statement, endorses the recognition of EMS by senior managers, usually understood as a public announcement of intentions and principles of environmental action. The adopted policy must describe the objectives of the organisation. This is not difficult if a company has made calculations. For example, the chemical industry decided to reduce emission by up to 95% over the next 5 years. The municipality may decide that after 3 years, the waste of 60% of the urban population shall be cleaned in a biological way.
- Environmental programme or action plan: It describes what actions the company will take next year. The environmental programme (action plan) transforms the company's environmental protection policy to goals and objectives, sets directions for activities, defines responsibilities of employees, and foresees human and financial resources. The industrial action plan should include the ways according to which each sector (workshop) should reduce emissions and what funds and instruments would be used to achieve the goal; also, the monitoring should be performed. Understandably, all actions should be coordinated by addressing these separate goals and one common goal of the policy. The programme is based on environmental aspects of the organisation, laws, and requirements, which should be taken into account. This information should be collected while preparing the initial environmental overview.

The organisational structure defines duties, authorises managers, and anticipates responsibility for actions. In that case, when a company is constituted of several branches with different activities, the organisational structure is framed both for the company as a whole and for its individual branches. The board member responsible for environmental protection reports directly to the director of the company. Individuals responsible for environmental protection strategy and organisation must be adequately provided with staff and financial resources.

The integration of environmental management into business is based on methods that integrate environmental issues into the functions of the enterprise, i.e. protection of employees, trading, research and development, product improvement, merger and acquisition of organisations, marketing, finance, and others. It consists of a variety of environmental procedures, usually contained in detailed production manuals and other working instructions describing the measures and actions needed to implement the environmental programme (action plan). Environmental procedures include

- information on important environmental issues, set out policy, objectives, tasks, and performed roles of employees in the EMS;
- information links within and outside of the company (with interested parties);
- EMS documentation and document control;
- labour control: procedures for operations and activities, criteria as well as goods, services, suppliers, and contractors of the organisations; and
- risk factors identification and response action plans for special cases, and protection against possible accidents.

Monitoring, measurement, and record-keeping procedures are designed to capture and monitor the results of specific actions or programmes, as well as the overall environmental impact. Correction and prevention actions eliminate possible and existing causes of deviation from objectives, criteria, tasks, and non-compliance with characteristics. During the EMS audit, the effectiveness and consistency of its implementation, as well as management review and formal assessment of the EMS adaptation to existing conditions prepared by the company's managers, are evaluated. Employee training contributes to the understanding of all employees why and how the additional duty to protect the environment is implemented to their functions. Public relations are necessary for it, and interested parties to inform about company's environmental objectives and activity provide necessary information on specific environmental issues, as well as difficulties and everything else that could influence interested parties.

4.5 Eco-Labelling of Goods

Since the EMS is an environmentally sound approach to the whole life cycle of a product (production factors, final product, and utilisation), labelling is also important to encourage consumers to choose a particular product or service. Eco-labelling is the most commonly used way of providing information to consumers about the environmental characteristics of products. The main objective of eco-labelling is to encourage a more balanced production and consumption of products.

The environmental label (EL) is a graphic symbol that appears on the label or the advertisement of the item. In this way, a person is informed about choosing environmentally friendly goods. Eco-labelling is the most commonly used way of providing information to consumers about the environmental characteristics of products. The main objective of eco-labelling is to promote a more balanced production and consumption of products.

Over the past years, the demand to reduce the adverse environmental impact of business activities has led to the need to assess the environmental impact of products from production to consumption (Santos 2007). According to Dagher and Itani (2014) and Scott and Vigar-ellis (2014), a key element of this product life-cycle chain is understanding consumer behaviour, which explains the decision-making process that determines how consumers buy, use, and dispose products.

Increasing the demand for organic products may, to a large extent, contribute to environmental well-being and must encourage companies to produce greener products. Although the environmental objective is to reduce the harmful environmental impact of consumption by producing organic products (Gallastegui 2002), from a business perspective, investors will have little interest in investing in organic products if consumers do not receive sufficient and accessible information about their products (Testa et al. 2015). One of the strategies that have been developed to eliminate the obstacles of the relationship between environmental protection and business is the use of "green labels", otherwise known as eco-labels.

Therefore, the use of "green labels" has three main goals:

1. to influence consumer behaviour, which would enable organic consumption;
2. as a marketing tool for companies to differentiate their products and attract "green" consumers, i.e. consumers concerned about the environmental impact of their lifestyle and consumption habits; and
3. to encourage companies to improve the environmental standards of their products.

According to Gallastegui, (2002), eco-labelling (the use of "green labels") affects consumer behaviour in two ways. First, this trademark introduces and initiates the concept of "green" as a characteristic that consumers must take into account when buying. Second, these trademarks allow/help consumers buy organic products in comparison with conventional products. In other words, it is easier for consumers to distinguish between organic and less organic products.

Eco-labelling is the most commonly used way of providing information to consumers about the environmental characteristics of products. The main objective of eco-labelling is to encourage a more balanced production and consumption of products.

Eco-labelling is a voluntary way of expressing environmental action. Labelling of organic products is widespread across the world. Eco-labelling of goods – products and services – is the environmental assessment of goods, when discussing the life cycle of products. The eco-label is granted to the product by a third party, which has fully independently assessed the compliance of the product with the ecological criteria. This distinguishes the eco-label from the so-called "green" labels, declared by the product manufacturer or service provider itself.

There are three groups of ELs, classified according to the criteria of ISO, which are included in the ISO 14020–14025 standards:

- ISO Type I – independent, third-party approved ELs that normally mark extensively used goods. Labelling programs are voluntary and provide products with labels that represent advantage and safety throughout the life cycle (for instance, the EU Ecolabel established by the European Community).
- ISO Type II – corporate environmental statements that usually inform about one aspect of the environmental impact of a product. Created by manufacturers, importers or distributors themselves. These labels are referred to as one-party labelling because the credibility of the label is not certified by an independent organisation (e.g. declaration by the manufacturer that the product contains x% of secondary raw materials).
- ISO Type III – environmental declarations of standard content products, providing quantitative product information in the form of indexes or indicators. Demonstrates the ecological value of the product over its life cycle. This type of eco-labelling enables comparison of product information.

The most common ISO eco-labelling is Type I. However, in some European Union countries (Sweden, Germany, Denmark, Norway), environmental product declarations (EPDs) are becoming more and more applicable.

Environmental product and service labelling programmes are based on the following principles (Ruževičius 2007):

- Voluntary participation of producers
- Uniformity of assessment criteria for eco-labels for domestic and imported products
- Clear formulation of eco-labelling criteria
- Compliance of the system with national laws and regulations
- Information is provided through the EL

In general, labelling provides information about the harmlessness of a product or service, informing the public when choosing more environmentally friendly products and thereby reducing the environmental pollution. Product eco-standards, the objectivity of compliance, and environmental education lead to the relationship between consumer and producer that help reduce environmental contamination (see Figure 4.5).

The concept of "organic" and "natural" cannot be equated. Labels such as "made from natural ingredients only", "hormone-free", and "non-genetically modified" do not indicate that the products are organic. The term "organic" is established in the European Union and Lithuanian legislation and that is certified in one of the EU certification bodies/authorities agricultural and food products, which must be grown, processed, labelled, and disposed of

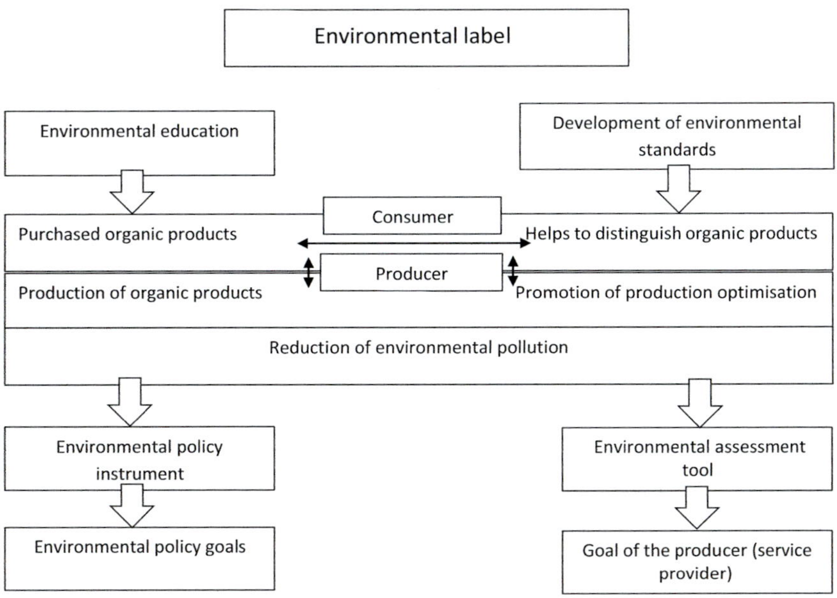

FIGURE 4.5
The environmental labelling system.

Source: Created by authors based on Ruževičius 2007.

in accordance with strict regulatory requirements, which do not contain synthetic materials, hormones, antibiotics, and genetically modified organisms.

The variety of logos of organic products implies the aim of controlling the overall EU organic production system (EU organic production logo) and promoting country-specific marketing techniques (Organic product label).

The labelling of organic products must include

- organic product identification logos;
- the code of the inspection body that carried out the last preparation operation;
- an indication of the origin of the raw materials of the product, showing where the raw materials contained in the product were grown; and
- a reference to the organic nature of the ingredients of agricultural origin.

These points on the labelling shall indicate that the product meets the requirements of organic production.

The most popular and commonly used logos are listed below.

Standard ISO 14000: ISO 14000 is not the real eco-label, but it indicates that the organisation has an EMS in place to reduce the environmental impact (http://www.iso.org/iso/iso_14000_essentials).

"The Blue Angel" is the world's oldest EL, launched in Germany back in 1978. It inspired other countries in Europe and beyond to follow the example of Germany and set up a national or regional labelling system for environmentally friendly products. This label covers several thousands of products and services: tires, construction materials from recycled glass, construction materials from recycled paper, carpeting, drinking water tanks, calculators, trash cans, solar cells, public transport, cell phones, handkerchiefs, gardening equipment, and many others (www.blauer-engel.de/en/).

The EU Ecolabel: The EU Ecolabel is awarded to products in 26 categories that meet high environmental standards throughout the product's life cycle: from production, packaging, transportation, use, and recycling (http://ec.europa.eu/environment/ecolabel/).

Former Current

Organic food producers in the European Union have been using the organic farming label since 1990. It is not used anymore since 1 July, 2010. However, you can still find it on some products in the supply chain.

Since July 2010, the European Union is required to display the new EU organic logo on all organic prepacked foods. Voluntary labelling of organic food produced in the European Union or imported from third-party countries on a voluntary basis is also possible.

Consumers buying products bearing the EU logo can be confident that

- at least 95% of the ingredients of the agricultural product are produced organically;

- the product complies with the requirements of the official control system;
- the product came directly from the manufacturer or processor in sealed packaging; and
- the product shall bear the name of the manufacturer, preparer or dealer and the name or code of the inspection body (http://ec.europa.eu/agriculture/organic/splash_en).

Organic farming: Lithuanian eco-friendly products can be recognised under this label. It is used by Lithuanian organic farmers and processors who have a document certified by the Lithuanian certification body "Ekoagra". The label means that the products are grown, processed, and marketed according to high standards.

Öko-Tex 100, 100+, 1000: Used since 1990. The world's most recognised eco-label for textile product groups. 8,000 manufacturers: 80 countries. 48.5% of certified products in the European market. 48.6% from Asia, the rest from America, Africa, Australia (http://www.oeko-tex.com).

Öko-Tex 100, 100+, 1000 means

- no paint that may cause allergies is used;
- the amount of pesticides and chlorinated phenols is being monitored;
- the amount of heavy metals content is monitored;
- no formaldehyde or traces of formaldehyde;
- no chlorine organic compounds;

- no biologically active finishing materials; and
- pH is suitable for the skin.

Energy Star: A label for electrical appliances that are energy and economically efficient. The "Energy Star" label is awarded to computers, electrical appliances, heating and cooling systems, and lighting equipment based on their energy consumption efficiency, durability, and quality (http://www.eu-energystar.org).

The EU Energy Label: The EU Energy Label provides a simple and clear indication of the energy efficiency of a device (from the best of class A+++ to the worst of class D) and several additional parameters (sound level, water consumption, and other.). At sale points, energy labels are placed on washing machines, refrigerators, microwaves, TVs, light bulbs, cooling equipment, and other appliances (http://www.newenergylabel.com).

Ozone free: This label indicates that the product does not contain ozone, a chemical that reacts very quickly with other chemicals and is hazardous to health, respiratory tract and causes respiratory diseases. Products containing ozone release negative ions into the environment, which attract various pollutants and accumulate in the air. And if ozone dissipates in the fresh air, indoors it accumulates on surfaces (furniture and others) (http://www.freedrinkingwater.com/general-health/air/ozone.php).

In conclusion, eco-labelling standards must

1. influence ecologically neutral consumers to start considering eco-friendly products, and
2. facilitate the behaviour of organic consumers in making the right choice of products.

Figure 4.6 provides a model of the impact of eco-labelling on environmental well-being through consumers' behaviour.

Around 80% of the US adult population state that purchasing more environmentally friendly products, such as other sustainable development activities, becomes the "new norm" when purchasing so-called "green products" (Report by the Shelton Group 2013). However, buyers are confused by many uncertain and sometimes misleading terms used to distinguish more environmentally friendly products in the market. This confusion can lead to even a sceptical view of the true meaning and value of more environmentally friendly products. If consumers might be misled about environmental requirements, the same might happen to businesses and organisations. Often, well-thought-out efforts to find a favourable position in the market for organic products are not combined with the principles of advertising

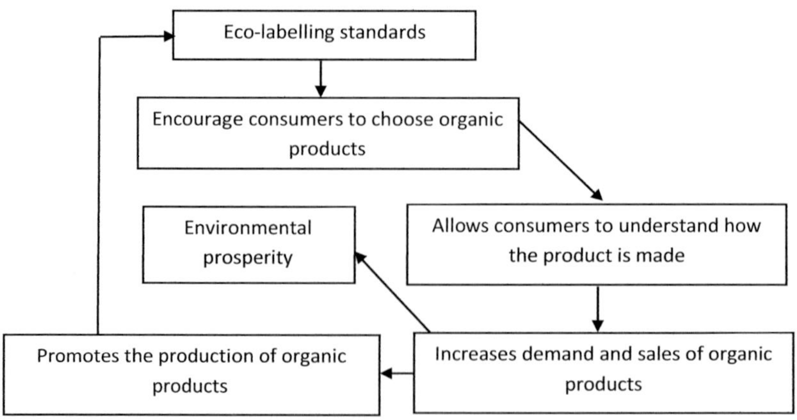

FIGURE 4.6
The positive impact of eco-labelling standards on consumer behaviour.

Source: Created by authors based on Ihemezie et al. 2018.

regulation and the requirements of truth-in-advertising. Only the provision of and dissemination of information on the environmental profile of a product throughout its life cycle can remedy the situation.

In general, more environmentally friendly products and services will have an increasing market share in the future. The US Department of Commerce has estimated that the US internal market trade turnover in so-called greener products and services exceeds USD 400 billion to USD 500 billion a year (Report by The Shelton Group 2013). And, despite the economic slowdown of 2008–2011 in the United States, growth rates in more eco-friendly market segments have outpaced the growth in similar conventional market segments, such as green products, green construction products, and renewable energy. Three out of four companies reported an increase in sales of more eco-friendly products during this period, and almost half (49%) reported an increase in sales of more eco-friendly products of 10% or more.

It is important to note that consumers can purchase organic products; however, if they use them not properly, it might not cause the intended effect of reducing the environmental impact of consumption (Ihemezie et al. 2018).

The EU Ecolabel makes it easy to know that a product or service is environmentally friendly and of good quality. It is easy to recognise, and it is reliable. The logo can be found on the packaging of each EU Ecolabel product:

To be awarded the EU Ecolabel, products must meet particularly strict criteria. These environmental criteria, established by the expert group of various interested parties, including consumer organisations and industry, cover the whole life cycle of a product, from the extraction of raw materials, the production, packaging, and transportation of the product till the end of the use of a product. In other words, the entire life cycle of a product is assessed.

Such an assessment of the life cycle of a product ensures that products bearing the EU Ecolabel have a lower environmental impact than similar products on the market.

Until March 2019, in the European Union, 1,575 licenses were issued to 72.797 products and services on the market. Thus, since 2016, the number of EU Ecolabel products/services has increased by 88%.

Changes in the number of eco-labelled products and services in the European Union from 2010 until 2019 are shown in Figure 4.7.

The fastest-growing groups of eco-labelled products are copying paper, paints and varnishes, and textiles. The leading countries in this area are Spain, Italy, Germany, Belgium, and France. Changes in licensing of organic products and services in the European Union from 200 until 2019 are shown in Figure 4.8.

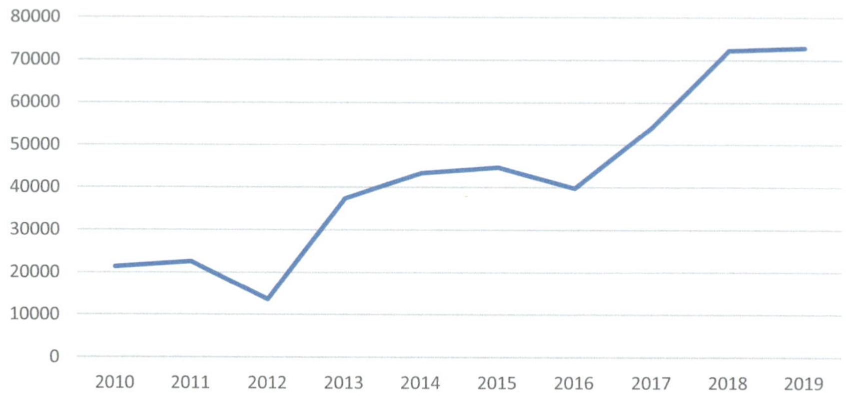

FIGURE 4.7
Changes in the number of eco-labelled products and services from 2010 until 2019 in the European Union.

Source: European Commission 2019.

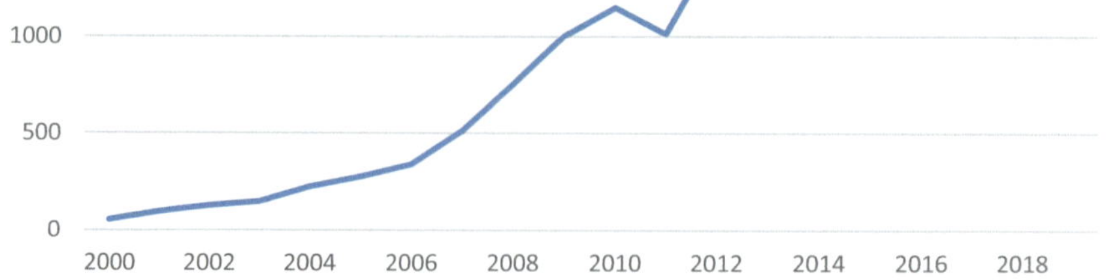

FIGURE 4.8
Changes in the licensing of organic products and services in the European Union from 2000 until 2019.

Source: European Commission 2019.

The decrease in licenses granted for organic products and services in 2019 is related to new criteria for companies wishing to obtain the Ecolabel. In 2018, the European Commission re-evaluated the criteria for the groups of eco-labelled products and services to ensure the best environmental protection efficiency and product quality.

Licences that have not yet been updated according to the new criteria are temporarily eliminated from the statistical data. The update process can take several months, companies have to demonstrate full compliance with the new criteria and carry out a new internal audit, and only at the end of the process, if the assessment is positive, their products/services will be integrated into the statistics.

The number of licenses in the EU countries as of March 2019 data is presented in Figure 4.9.

In March 2019, the highest number of licenses was granted in France (19.5%), Germany (18%), and Spain (12%). At the same time, Spain (42.5%), Italy (12%), and Germany (6%) had the largest market for organic products and services (see Figure 4.10).

In March 2019, there were 234 products and services bearing the EU Ecolabel in the Lithuanian market, while in Spain, there were 30,877 products and services (see Figure 4.10).

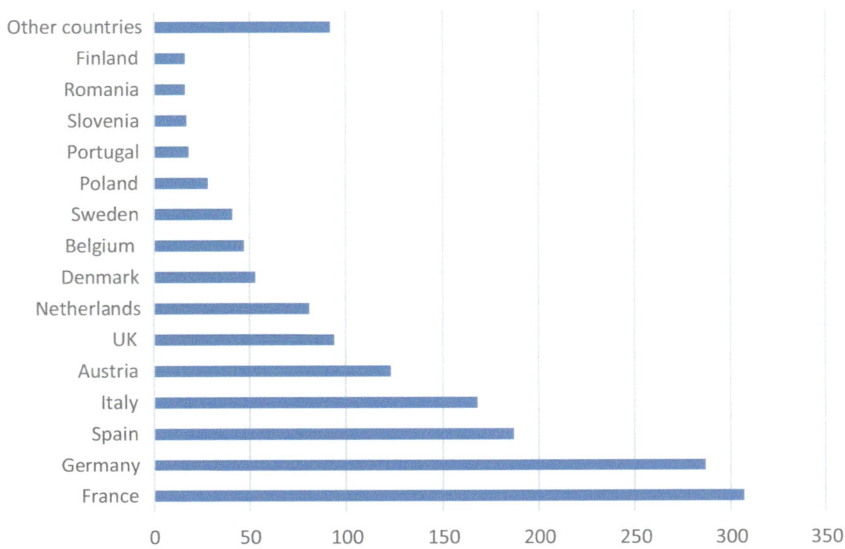

FIGURE 4.9
Changes in licenses for organic products and services in EU countries.

Source: European Commission 2019.

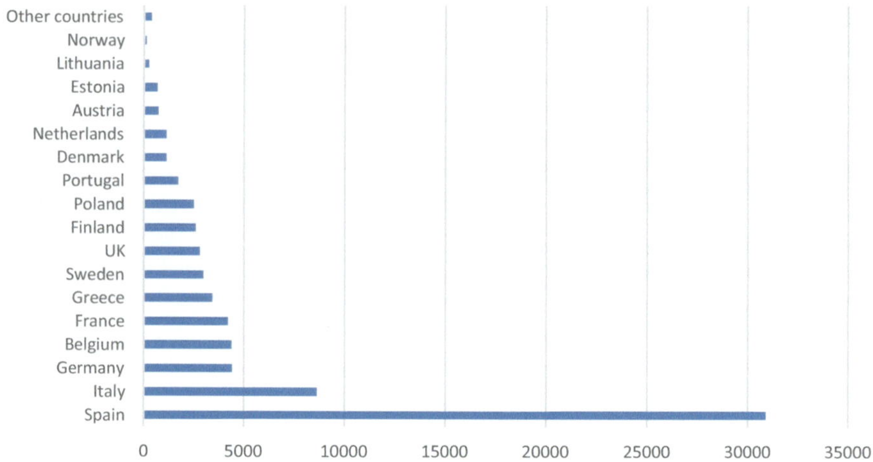

FIGURE 4.10
Number of eco-labelled products and services in EU countries.

Source: European Commission 2019.

5

Adaptation to Climate Change Impacts

5.1 Climate Change and Migration

In the general sense, migration (lot. *migratio* – moving, travelling) is the mobility of people determined by a variety of factors. Many authors have tried to form the best definition of migration.

Different approaches to migration have led to different definitions of it as a phenomenon:

- Migration (lot. *migratio* – moving, travelling), moving, travelling, advancing from one location to another; relocation of residents from one place to another (Dictionary of International Terms).
- Migration is the movement of people into the country (immigration) and from a country (emigration) (Dictionary of Economic Terms).
- Migration is the movement of people from one place of residence to another, by crossing the administrative boundaries of a location (Papademetriou and Martin 1991).
- Migration is a change of the place of residence of people, the main result of which is a change in the number and structure of the population both in the country of origin and in the country of destination (Brazauskienė and Kazlauskienė 2002).
- Migration is defined as a departure from a country or arrival in a country from another foreign state in pursuit of a better quality of life (Urbonaitė 2007).
- Economic migration is defined as a departure from Lithuania with the goal of settling in a foreign country (i.e. economic emigration) or arrival from a foreign country with the goal of settling in Lithuania (i.e. economic immigration) in pursuit of a better quality of life, higher wage, and better working conditions (Economic Migration Regulation Strategy of the Republic of Lithuania 2007).

From different definitions, two trends can be observed: migration can be defined in a narrow and broad sense. In a narrow sense, migration is defined as the movement of people from one location to another (Stockwell and Groat 1984). In a broad sense, possible causes and consequences are presented, for example, the definition by Brazauskienė and Kazlauskienė (2002) identifies the evolution of the country's population as the result of migration (both from the perspective of the country of origin and the country of destination). To summarise the definitions given by different authors, it may be stated that migration is a departure or arrival in another country in pursuit of personal economic prosperity.

In scientific literature, the aspect of migration in the sense of time is also emphasised. Migration is divided into long-term (from one year) and short-term (lasting from 3 months to 1 yearlong). According to Damulienė (2013), in terms of the economy of the country of origin, long-term migration is more harmful, as many of those who leave for long periods of time usually do not return, investments into education are lost, and ultimately specialist of the specialist field are given up (brain drain). Thus, it can be argued that there are two important dimensions in the search for a unified definition – dimensions of space (relocation, movement to live elsewhere) and time (long-term, short-term). It is, therefore, appropriate to update the previously mentioned definition with the dimension of time: migration is a long-term departure or arrival in another country in pursuit of personal economic prosperity.

Scientific literature distinguishes separate types of migration, depending on the various aspects that determine migration. Most often, migration is divided into separate groups:

- Direction: In terms of direction, migration can be internal and external. Internal migration is confined within a country's borders and does not cross them, for example, relocation of individuals from rural areas to urban areas and vice versa. In contrast, external migration takes place on a global scale, as it is a relocation from one country into another, resulting in a change in the composition and the level of the population of both countries involved in the migration.
- Departure/arrival: This aspect encompasses immigration (arrival in a country), emigration (departure from a country), re-emigration (return of emigrants), and also, as defined in the post-war period, repatriation (return of citizens [exiles, captives, diasporas] to their homeland.
- Duration: In terms of duration, migration is divided into four subgroups: permanent (long-term), temporary, seasonal, and daily. Permanent migration occurs when an individual relocates to another country for a long period of time and settles there permanently. Temporary migration involves a short period of time, although in literature the short term is not precisely defined, it is estimated that it can last from 1 day to several years. This is a temporary relocation to live in another country with the goal of returning. Seasonal migration is very popular among students and young people.

- It is the migration for a certain season, for example, for agricultural summer work. Daily migration is not a common phenomenon, as it lasts less than 24 hours but occurs every day.
- Format. By format, migration is divided into voluntary and forced. Voluntary migration is carried out by an individual of his own choice, whereas forced migration encompasses the use of force and when individuals have no choice, for example, war, natural disasters, religious persecution, etc.

There are three types of migration:

- Forced migration – the decision to migrate is made by other individuals, irrespective of the person's willingness. Most often it is politically motivated migration (repression, war).
- Labour force migration – economically dictated. Migration to improve economic opportunities and living conditions.
- Family migration – an integral part of labour force migration and forced migration. Family members come to the migrants.

Bell et al. (2010), having examined the causes of migration, have identified other types of migration that manifested notably in recent decades. The authors divided migration into three main groups: labour force migration, forced migration, and international retirement migration. Labour force migration is divided into the migration of qualified and unqualified labour force. The migration of qualified labour force is highly welcomed and even encouraged by the accepting country; however, from the point of view of the country of origin, this type of migration promotes "brain drain". The majority of the attractive countries, such as the United States, Canada, or Australia, have a certain point system for classifying and rating migrants who express a wish to enter the country. However, despite the favourability, this type of migration represents a very small percentage of migration as a whole. The amount of unqualified labour is the highest as usual; it is cheap labour from economically underdeveloped or developing countries. A great example is the European Union and the free movement of labour when attractive countries like the UK or Ireland are overcrowded with migrants, and as a result, social issues and migrant support with state funds are encountered. Also, this type of migration causes problems in the country of destination and is mostly illegal (the case of the United States and Mexico). Forced migration, according to the authors, includes not only refugees and asylum seekers fleeing war or repression but also people migrating due to projects, such as dam construction, or natural disasters. Migration due to retirement is a phenomenon in the Northern EU members, such as Germany, the United Kingdom, or the Nordic countries. The best characterising factor for this type of migration is retired individuals who have enough savings to be able to purchase property in a foreign country. The driving factor, in this case, is retirement, and the attraction factor is a favourable climate. The benefits of this migration are most noticeable and felt in Portugal and Spain, where the mass construction on the coasts is aimed specifically at the migrant market.

Research of migration as a phenomenon only started in the 1980s; however, as its scale keeps growing massively and it is becoming a global phenomenon, the interest in this process has been increasing as well. A lot of research has done to define the factors that cause this phenomenon. In the explanation of these factors, the main theories of migration have been distinguished (Table 5.1).

The theories presented in Table 5.1 explain the causes of migration and the factors that result in migration. General factors that cause migration of the population can be distinguished from the theories:

TABLE 5.1

The Theories and Factors of Migration

Theory	Factors
The neoclassical economic theory	Influenced by the factors of pushing and pulling. The pushing factor is a poorly paid job. The pulling factor is the sale of one's labour for a greater income in newly emerging markets.
The relations theory	There are migrant networks and interpersonal links between current, former, and future migrants. Migration is likely because it has lower settlement costs.
The segmented labour market theory	The labour market consists of two segments. The first one (or the upper one) is a segment (labour market) of educated, high-income earners, and the second one (the lower one) is the poorly paid segment (labour market) of local unfilled jobs is filled up by immigrants.
The new economic migration theory	Explored at the level of the family and household, not at the level of the individual. Migration is a rational choice and also a strategy of a family/group. Under difficult economic conditions, part of the group's employees are relocated abroad to maintain the economic stability of the household.
The world systems theory	The world's labour potential is divided into three zones: the main one, the semi-peripheral, and the peripheral. The labour force moves from the peripheral zone (developing countries) to the main zone (rich capitalist post-industrial countries). This is the migration of cheap labour into stronger countries where the larger wage is paid for the same unqualified labour.
The neoclassical macroeconomics theory	In society, a balance must be reached between the two main economic resources: labour and capital. The labour force is moving into larges wage countries, while the invested capital will move into low wage countries to maximise profit.
The neoclassical microeconomics theory	Individuals make conscious decisions on migration issues. If the profit after the relocation into another country is expected to be positive, individuals tend to migrate. When calculating the expected profit, the following is considered: the probability of getting a job, the material and psychological expenses of relocation, and in case of illegal work – the possibility of being tracked down and deported.

Source: Created by authors based on Douglas et al. (2011).

- Economic (wage, economic stability, unemployment, and others)
- Demographic (marriages)
- Social (traditions, intolerance, culture)
- Political (conflicts, repressions, liberalisation of migration, election results)
- Psychological (personal decisions)
- Geographical (climatic conditions)

All theories presented in Table 5.1 seek to explain the origin of migration and the factors that cause it. The theories do not contradict one another unless a theory accepts the factor of migration as the only one and functioning at a single level. However, most often a combination of these factors is accepted. It is stated that individuals count the economic benefit, households diversify the division of labour, and the socio-economic context, in which decisions are made, is determined by the present national and international forces (Papademetriou and Martin 1991).

According to Damulienė (2013), who researched migration theories, an individual, not exposed to any additional forces, will base his decision to migrate on an assessment of the benefits and drawbacks of the two locations (countries) – a comparison of the destination country and the departure country. From the theories given in Table 5.1 and Damuliene's insights, it can be concluded that migration is caused by two main factors (this is best explained by the neoclassical economic theory):

- pushing (push factors), from the location of residence, and
- pulling (pull factors), into the location where one wishes to live. Individuals migrate to where pull factors are functioning.

Čiarnienė and Kumpikaitė have grouped the factors that encourage migration into four groups (Figure 5.1). In addition to the mentioned pull factors (researchers use the term "demand") and push factors ("supply"), the researchers distinguish chain reaction factors and psychological factors.

Demand factors (pulling) include relocating to a country that is economically stronger than the country where the individual lives. The factors assigned to this group include more favourable tax system, higher average wage, larger labour market, better education system, more developed healthcare system, political stability, and greater tolerance. Supply factors (pushing) are related to the internal situation of the country in which the individual currently lives. Possible factors include war or other forms of armed conflict in the country, poverty, unemployment, low average wages, high social exclusion, absence of the middle class, and poor political situation in the country. The third group of factors is chain migration. Individuals who have relocated earlier inform the remaining people about the living conditions and the standards of living provide the

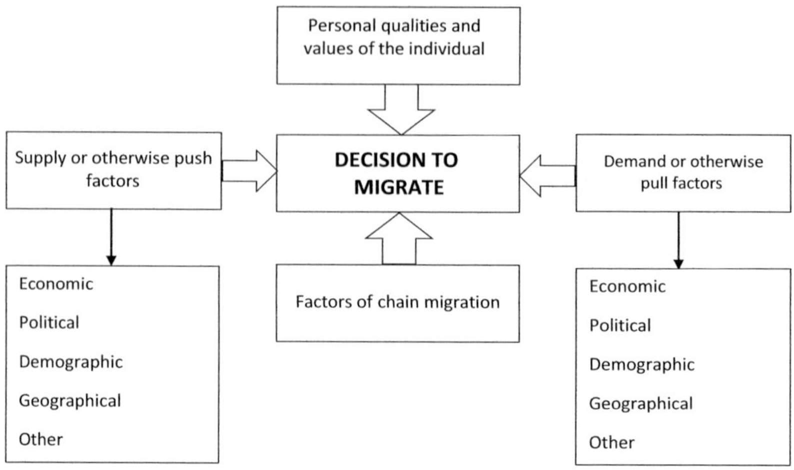

FIGURE 5.1
Factors of the decision to migrate.

Source: Čiarnienė and Kumpikaitė 2011.

main initial information about the accommodation, work, transportation, and arrival. Consequently, individuals evaluate the feedback and make a decision, thus causing a wave of emigration. The final group of factors is the personal qualities and values of the individual (psychological factors). The main links that bind the individual to the homeland are emotional relations with parents, relatives, friends, motivation, and language barriers.

Scientific literature also emphasises three levels of migration promoting factors: micro-level, macro-level, and mezzo-level. To the first level includes factors related to the individual's personal decision, motives, for example, education and language. The second level covers the labour demand and supply, and the wage differences in the destination country. The third level includes the personal environment of the individual – family and relatives. According to Čiarnienė et al. (2009), a person's decision regarding migration is most influenced by economic factors.

Also, a new factor for promoting migration processes has emerged in the modern world – globalisation. According to Rudžinskienė and Paulauskaitė (2014), globalisation as a phenomenon does not have a direct impact to migration, but it facilitates the migration process through the free movement of persons and technological development which allows them to communicate with the entire world. Thus, globalisation can be recognised as one of the factors that facilitate the migration process. Nevertheless, at the same time, globalisation can play the role of a suppressor of migration in individual cases, for example, companies hold off migration by creating new jobs in the countries of cheaper labour. Then another phenomenon is noticed – "arrival" of the job to the employees, which results in the absence of the main need for migration: search of a job.

From 1990 to 1 January 2019, 999,865 residents have already emigrated from Lithuania. During the time, 289,834 immigrated; therefore, the total foreign migration balance for this period is –710,031 residents. Net migration per 1,000 residents in 2015 is presented in Figure 5.2.

The net migration rate is the difference between immigration into and emigration from the area during the year, divided (usually) per 1,000 inhabitants for comparison reasons. Net migration rate has a positive value in a country that experiences more people entering than are leaving it, for example, the United Kingdom, Sweden, and Germany. Meanwhile, countries such as Lithuania or Latvia, where more people are leaving than entering, have a negative net migration rate.

In 2017, more than 76% of emigrants were aged between 15 and 44. This age group only makes up about 36% of the entire Lithuanian population. In the long term, as the number of young people is decreasing due to emigration, this causes serious demographic problems – diminishing number of marriages, decreasing birth rate, lack of workforce. A shrinking proportion of working people must support a growing number of the unemployed.

The reverse trend is seen when comparing the number of older emigrants to the population of Lithuania. People who are 65 or older only comprise about 1.5% of the total emigration flow, despite making up about 19% of the entire population.

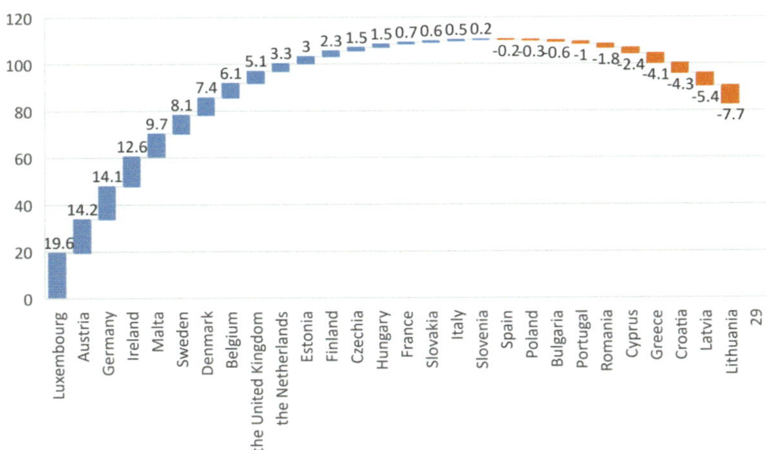

FIGURE 5.2
Net migration per 1,000 residents.

Source: Migration trends http://123.emn.lt/#chart-4-desc.

Emigration and immigration in Lithuania is happening continuously. Two major waves of emigration were observed, in 2005, when emigration increased by 53% compared with 2004 (from 37,691 emigrants to 57,885 emigrants) and in 2010, when the growth, compared with 2009, was 116% (from 38,500 emigrants to 83,157 emigrants). Lithuania's accession to the European Union and the 2008 financial crisis could be considered as a reason for emigration growth. In 2015, emigration started to grow again and increased by 22% compared with 2014 (from 36,621 emigrants to 44,533 emigrants). Immigration figures have been growing over the past decade. From 2005 to 2015, immigration to Lithuania increased by 226% (from 6,789 to 22,130 immigrants). Over the observed decade, two periods of decline of the immigration number can be seen: compared with 2008, in 2009 the number of immigrants decreased by 43% (from 9,297 to 6,487 immigrants) and in 2015 when the number of immigrants decreased by 9% (from 24,294 to 22,130 immigrants). The reasons behind these numbers of immigrants are identical to the cause of the wave of emigration growth during the period in question. Emigration significantly overtakes immigration. For example, the 2015 migration saldo is negative and reaches –22,403. However, in 2014, there were 8.3 immigrants per thousand of the Lithuanian population. According to the published statistics, in 2014, labour immigrants accounted for 22% of all foreigners in Lithuania.

International migration since 2000 is rapidly increasing (Figure 5.3). One of the factors contributing to the rapid rate of migration is climate change and phenomena related not only with nature but also in the labour market.

Climate change is drastically changing our world and is leading to unprecedented movements of people from one place of residence to another in order to reduce the growing level of human insecurity. We live in a global era of unprecedented movement of ideas, money, and, increasingly, people.

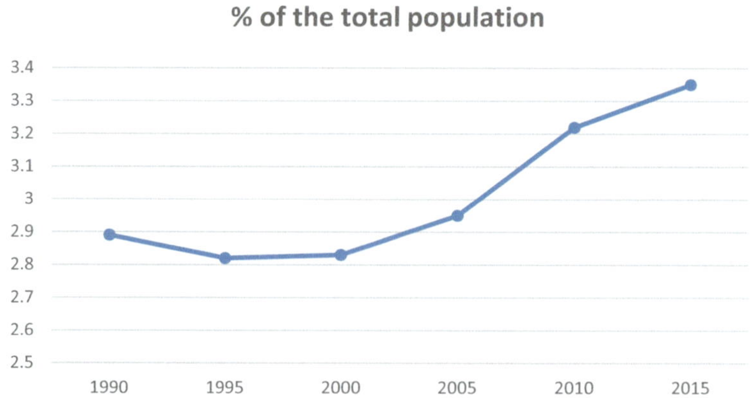

FIGURE 5.3
International migration.

Source: https://data.worldbank.org/indicator/SM.POP.TOTL.ZS.

The size of the human population, along with the way we consume resources, fundamentally changes our world. Even though our economic model "take – dispose of" has created prosperity for hundreds of millions of people in many countries, significantly reducing global poverty, it has also left too much waste and a too big environmental impact. Above all, future generations face huge social, economic, and environmental risks. Perhaps the greatest risk comes from filling the atmosphere with greenhouse gases, the amount of which is higher than at any time in the last 66 million years.

Today, one billion people are migrants, who have moved across the borders of their national states or even further beyond them. They moved for a variety of complex reasons, including population insecurity, lack of economic opportunities, environmental degradation, and new forms of travel. At the same time, these factors contribute to the resettlement of people and insecure migration. And both of these types of migration will only grow, as the impact of climate change will gradually reduce the security and livelihood of millions of people.

Climate change is fundamentally changing locations where people can live. Food supply is disrupted and insecure in the Sahel region in North Africa and Central America; water shortage is observed in North Africa and the Middle East. Somalia, for example, is experiencing more droughts. Iraq is struggling with more frequent heatwaves. Unexpected storms and floods have put residents of the Caribbean and the Gulf of Mexico under real threat.

Nevertheless, positive tendencies may also be observed. Firstly, we are making great progress in developing resistance to extreme weather conditions. In the 1970s, Bangladesh lost hundreds of thousands of people due to floods. Today, deaths due to similar events, although no less tragic, are much rarer, as we are better prepared for this kind of disaster. Secondly, for the first time in history, the international community is working together to form a framework for managing international migration. Intergovernmental negotiations began in February 2018, for the purposes of adopting a Global Compact for Safe, Uncluttered and Regular Migration (GCM). At the end of the same year, the United Nations General Assembly concluded the process and adopted the GCM.

This compact promises to provide a solid basis for actions to reduce climate change. The GCM is an opportunity of the generation to develop an internationally harmonised system for managing safe and uncluttered migration. Therefore, it can improve the lives and prospects of tens of millions of people.

The global compact on migration is the first global agreement of the UN on a common approach to international migration and all its aspects. The global compact version is not legally binding. It is based on state sovereignty, shared responsibility, non-discrimination and human rights values, and recognises that in order to optimise the migration benefits, it is important to cooperate and address the risks and challenges of individuals and communities in countries involved in the migration processes.

The general plan includes 23 goals on how to better manage migration at local, national, regional, and global levels.

Global compact

- aims to reduce the negative and structural factors that prevent people from creating and maintaining sustainable livelihood in their countries of origin;
- intends to reduce the risks and vulnerabilities faced by migrants at various stages of migration by respecting, defending, and executing their human rights;
- seeks to address the legitimate concerns of states and communities, while acknowledging that society is undergoing various demographic, economic, social, and environmental changes that may affect migration;
- attempts to create favourable conditions to enable all migrants to enrich our society through its human, economic, and social capabilities, which would facilitate their contribution to sustainable development at local, national, regional, and global levels.

Lastly, and most importantly, we will have to do everything possible to reduce greenhouse gas emissions significantly. This is the only way to keep Earth's temperature increasing not more than 2°C compared with the industrial level/threshold at which spiral feedback loops can cause recurrent climate changes.

Cross-border migration and its consequences for host countries are creating great concern and a lot of attention is paid to it. However, it is more and more recognised that many people migrate within their own countries than beyond national borders. They are moving for many reasons – economic, social, political, and environmental. Climate change has now become a strong driver of internal migration, encouraging more and more people to move out of vulnerable and not viable areas of their countries in order to create a new, safer, and better life.

Recently, the world is facing a new type of migrants, the so-called "environmental migrants", who are defined as "persons or groups of persons who, predominantly for reasons of sudden or progressive change in the environment that adversely affects their lives or living conditions, are obliged to leave their habitual homes or choose to do so, either temporarily or permanently, and who move either within their country or abroad" (Migration, Environment and Climate Change: Evidence for Policy (MECLEP) 2014).

Environmentally displaced persons mean "persons who are displaced within their country of habitual residence or who have crossed an international border and for whom environmental degradation, deterioration or destruction is a major cause of their displacement, although not necessarily the sole one" (MECLEP 2014).

Term "disaster displacement" refers to situations where people are forced to leave their homes or places of habitual residence to avoid the impact of natural hazards. Such displacement may be voluntary or ordered by the authorities, or by the form of evacuation. Such displacement may take place within the country or beyond international borders. Evacuation is the rapid movement of people away from an immediate threat or impact of a disaster to a safer place of shelter. It is commonly characterised by a short time frame, from hours to weeks, within which emergency procedures need to be enacted in order to save lives and minimise exposure to harm.

Evacuations can be:

- Mandatory: An evacuation ordered and directed by authorities when it is judged that the risk to a population is too great to allow them to remain where they are, and where sheltering in place would likely entail a higher level of risk. This places a duty of responsibility on authorities to ensure that people have the information and assistance needed for safe and timely evacuation and that evacuees are cared for.
- Advised: An official evacuation advisory message may be issued to enable early response and informed decision-making by the population at risk on whether and when to evacuate. An advisory may precede a mandatory order to evacuate as the level of the threat and the risk associated with the alternative of sheltering in place increases. As for mandatory evacuations, authorities are usually seen to have a responsibility to facilitate safe and timely evacuations for those in need of assistance.
- Spontaneous: When people evacuate their current location due to actual or perceived risk using their own means (self-evacuation) and without (or before) being officially advised or directed to do so. This may include people who leave areas outside a designated evacuation zone, also known as "shadow evacuations" (International Organization for Migration (IOM) 2014).

In 2018, 17.2 million people in 144 countries and territories were newly displaced within their own country in the context of disasters. The Internal Displacement Monitoring Centre (IDMC) records that worldwide, over a period of 11 years (2008–2018), about 265.3 million South and East Asia, and Pacific region residents were displaced to safer areas, and these regions were recognised as the most affected ones. In particular, the Philippines, China, and India recorded the most disasters in 2018.

Small island developing states (SIDS) have been disproportionately affected by natural hazards. In 2018, the displacement of the population has been caused primarily due to extreme weather events, especially storms (9.3 million people) and cyclones, hurricanes, and typhoons (7.9 million people).

The southwest monsoons in India, typhoon Mangkhut in China, and the Philippines were the most devastating (Global Report on Internal Displacement 2019). Although the majority of mobility in the context of climate change, including disaster displacement, takes place within the country, some people are forced to move abroad.

By 2050, an increasing impact of climate change in three densely populated regions of the world (sub-Saharan Africa, South Asia, and Latin America) can force more than 140 million people to leave their homes, and this is creating an impeding human crisis and threatens the development process of the whole world. However, with coordination of actions, including global efforts to reduce greenhouse gas emissions and strong development planning at the country level, this worst-case scenario, in excess of 140 million migrating people, can be reduced even by 80% or more than 100 million people.

It is very difficult to distinguish when the environment is the main factor of migration. In many cases, the environmental factors are closely linked with the social and economic, political, demographical, cultural, and personal factors, which play an important role in the mobility prevention (Laczko and Aghazarm 2009; Migration and Global Environmental Change 2014). There is still a lack of methodological measures for collecting and processing information on people migrating due to more gradual, so-called slow-onset impacts of climate change, for example, sea-level rise or salinity increase. It is not easy to assess the migration nature and extent, which is directly linked to environmental changes.

The decision to migrate is difficult. Impact factors change over time and vary between individuals and larger households. Climate change forces to make a difficult decision when/whether it is time to settle in another place and leave home.

In sub-Saharan Africa, South Asia, and Latin America, which together account for 55% of the world's developing countries population, it is observed that climate change will encourage tens of millions of people to migrate within their countries by 2050. In addition, without specific actions on climate change and development, it is estimated that just over 143 million people or about 2.8% of these three regions could be forced to move in their countries to avoid the increasing impact of climate change (Rigaud et al. 2018).

In a year, about 9 million people migrate between countries (Abel and Sander 2014) and about 244–250 million people live outside their country of origin (UNDESA 2015). Although the absolute numbers of today are higher than in the past, since 1990, international migration has not changed much (Guy and Sander 2014). More than 60% of international migrants (150 million people) are recognised as migrating workers, whose main reason for migration is the economic instability of the country.

Climatologists around the world agree that the question "will we experience climate change in the future?" turned into a question "how will we cope with the catastrophic effects of climate change in the future?"

5.2 Climate Change and Developing Countries

Despite international efforts, poverty has become more prevalent in many countries over the past decade, making poverty reduction a key task for the development in the 21st century. By signing the Millennium Declaration, 189 countries have decided to halve extreme poverty in the world by 2015, and all countries involved with this document are committed to contributing to this goal. Climate change poses a serious threat to poor countries around the world, and it is not clear whether the economic growth model developed over the past decades is suitable for the global world. As stated in the Johannesburg Declaration on Sustainable Development, "the adverse effects of climate change are already evident, natural disasters are more frequent and more devastating and developing countries more vulnerable". Although climate change is a global phenomenon, its negative effects are more evident in fragile economies, as these countries are more vulnerable due to their high dependence on natural resources and the limited capacity to cope with climate change and extreme conditions. The poorest, with the least resources and opportunities for adapting, are the most vulnerable in these countries.

The greatest challenge to adapt to climate change is in developing countries (e.g. Tol et al. 2004; Mendelsohn et al. 2006) for the following three reasons:

- The first reason is geography. Many developing countries are located in tropical and subtropical regions and therefore are already in hotter climate conditions where it is more difficult to develop certain economic activities. Further increase in temperature will lead to even less optimal conditions (Mendelsohn and Schlesinger 1999). Certainly, climate change is more than just temperature, and many of the effects of climate change are expected to be related to potential changes in access to clean water. In this framework, developing countries, which are generally located in geographical areas with low to moderate rainfall, once again face the greatest threats.
- The second reason is often referred to as sensitivity. A relatively high proportion of developing countries' production comes from sectors which are particularly vulnerable to climatic conditions, mainly agriculture. In general, the concept of sensitivity can include the fact that many people in developing countries are already using marginal livelihoods.
- The third reason is the lack of adaptive capacity. Developing countries often lack resources to adapt to climate change, including financial resources (both, savings and access to credit), good governance, infrastructure, and information.

Expected changes in the frequency, intensity, and duration of the extreme weather conditions (such as heatwaves, heavy rainfall, and droughts), as well as more gradual changes in the temperate climate, pose a particular threat to their livelihoods, leading to further increase in inequality between developing and developed worlds. This reaffirms that climate change is a serious threat to poverty eradication. However, current development strategies tend to overlook the risks of climate change.

The researchers recommend a method that combines both mitigation and adaptation. Current commitments to mitigate climate change by limiting greenhouse gas (GHG) emissions, even if not implemented, will stabilise atmospheric concentrations of these gases. Capacity to adapt must be strengthened to reduce the losses of climate change mitigation losses and rapidly consolidate climate change mitigation efforts.

Therefore, adaptation to climate change is an appropriate response to climate conditions, which reduce vulnerability, and is an integral and urgent part of overall poverty reduction strategies. Adaptation should not be seen as a stand-alone activity isolated from other environmental and socio-economic challenges that also affect the development opportunities of poor countries. A comprehensive approach is needed that would take into account the potential synergistic and antagonistic effects between local and global environmental changes, as well as socio-economic factors.

Adaptation to climate change is worth a broader exploration. The Intergovernmental Panel on Climate Change defined it as "any adjustment in natural or human systems in response to actual or expected climatic stimuli or their effects which moderates harm or exploits beneficial opportunities" (Smit 2016). It is important to distinguish between the concepts of "flow" and "stock". Flow adaptation includes adjustments that result in both costs and benefits over a single period of time (such as changes in variable agricultural costs, crop varieties and fertilisers, and changes in indoor heating and cooling). On the contrary, stock adaptation is a form of investment where the costs are paid off and the benefits are gained over several periods of time in the future, for example, dams protecting coastal populations from rising water, or dams that reserve water to protect areas from drought. However, stock adaptation may also include other, indirect forms, such as investing in healthcare infrastructure to cope with the burden of evolving diseases due to climate change or investing in agricultural development programmes to support farmers' adaptation to changing conditions due to climate change.

Two schools can be distinguished related to the best adaptation strategy in the developing world:

- The first one states that "development is the best form of adaptation": It is better to prioritise traditional development goals, i.e. invest in physical and human resources reserves and robust institutions to protect investments specifically designed to reduce vulnerability caused by climate change. The logical basis for this statement might be Thomas Schelling (1992), who reasoned that, in view of developing countries that are vulnerable to climate change, because of their high sensitivity and low adaptability, which are essentially minor problems for them, the best protection against climate change for these countries can only be their continuous development (Mendelsohn 2012). It is important to note that Schelling was thinking about whether adaptation rather than emission reduction was the best policy response to climate change. Nonetheless, his suggestion led to further work, directly arguing that development is the best form of adaptation (Mendelsohn 2012; Fankhauser and Burton 2011).

- The second school of thought claims that "development depends on adaptation": Climate change will pose a serious threat to the development process unless specific adaptation takes place. In this regard, successful defensive adjustment is a necessary prerequisite for building capital reserves and improving development-related prosperity. This concerns the political stance taken by various international development organisations.

There is a lot of literature on adaptation to the effects of climate change, but much of it is based on local research. Only a few studies have shown that adaptation is a macroeconomic issue. Their main objective was to quantify the cost of adaptation at the national and regional levels, supporting international political negotiations to reach an agreement on the adjustment of industrialised and developing countries to the changed payments. Fankhauser (2013) discerns this literature into "first generation" and "second generation".

First-generation studies were very simple. Their approach simply aimed to estimate the proportion of current climate-dependent investments and then multiply them by the corporate tax rate increase, which represents the total value of "climate-resilient" expenditures (Stern 2007; World Bank 2006). This method was static, and in particular, the increased coefficient had no foundation which would be based on empirical research.

Second generation studies were launched under the United Nations Framework Convention on Climate Change (UNFCCC) guidelines (2007), which detailed the costs of adaptation in key climate-sensitive sectors. It had a stronger empirical basis but remained static. Research published by the World Bank in 2010 estimates the costs of adaptation but critically limits the assumption that adaptation occurs until all climate damage is eliminated, which cannot be effective. Indeed, this is the main criticism of all these studies, which states that development is the best form of adaptation (Mendelsohn 2012).

Several recent studies have expanded the "integrated assessment models" of a linked climate economy system by including a control variable. Previously, these models did not consider adaptation since it was indirectly associated with damage to the climate, and therefore the planner could not change it. Adaptation-integrated assessment models incorporate a dynamic view of adaptation and quantify its benefits. The use of such models has confirmed that optimal adaptation leaves "residual" losses due to climate change and that optimal climate policy includes both adaptation and emission reduction.

Recent and projected aspects of regional climate change impacts (IPCC 2014a–g):

Africa

- Africa might be the most vulnerable continent to climate change and change due to many existing political and social issues and low adaptive capacity. Existing problems include poverty, unsafe food, political conflicts, and the degradation of ecosystems.
- It is estimated that by 2050 due to climate change, 350–600 million people will face a shortage of water.
- The urban population is projected to triple in size (to 800 million people), which will lead to even greater urban poverty, obstruction, or even eradication of access to basic services in the cities.
- In many African countries and regions, agriculture is expected to be severely affected by climate changes, which will jeopardise access to proper nutrition.
- At the end of the 21st century, projected sea-level rise is likely to affect low-lying coastal areas with large populations, including Senegal, Liberia, and Mozambique.
- Climate variability and pre-existing changes can have a negative impact on human health. In many African countries, climate change may increase existing health threats, such as malnutrition, malaria, and other diseases.

Asia

- Asian glaciers are melting faster than ever according to historical records. Currently, some glaciers account for 20%. Melting glaciers increase the risk of floods and snow avalanches from destabilised slopes.
- It is predicted that there will be limited access to freshwater due to climate change, especially in Central and Southeast Asia, and particularly in large river basins. The steadily growing population and increase in needs related to population growth could negatively affect more than 1 billion people by 2050.
- Coastal areas, especially the densely populated deltas of the South and Southeast Asia, will be threatened by the increase in the number of sea floods (sometimes including river floods).
- The effects of climate change on crop yields can vary greatly depending on region, crop type, and regional changes in temperature and precipitation. By the mid-21st century, for example, climate change in

East and Southeast Asia can increase crop yields by up to 20% and in Central and South Asia by up to 30%.
- Deaths from diarrheal diseases are likely to increase in East, South, and Southeast Asia due to the expected changes in the hydrological cycle associated with climate change.

Australia and New Zealand

- With global warming reaching 1°C, water security concerns in the southwest and southeastern Australia and in the northern and some eastern parts of New Zealand are expected to become major.
- Changing biodiversity in some ecologically rich areas, including the Great Barrier Reef and the Wet Tropics of Queensland, by 2050.
- Sea-level rise, more powerful storms, and coastal floods will continue to affect coastal areas. Coastal development and population growth in areas such as Cairns and South East Queensland (Australia), and Northland and Bay of Plenty (New Zealand) pose a threat to many people and infrastructure.
- Increased numbers of droughts and fires are expected to reduce agricultural and forestry production in much of South Australia and northern and eastern parts of New Zealand.
- Cascading and interacting economic, social, and daily life in rural areas can lead to long-term drought, which is affecting the survival of certain regions. The prolonged drought has been linked to increased suicide rates among male farmers in Australia.
- Extreme storms can increase the likelihood of embankment, dams, drainage, and sewer systems failure. They can also increase the damage caused by storms and fires.
- More frequent heatwaves can lead to an increased number of deaths and more electrical disturbance.

Europe

- Europe is already documenting the widespread effects of climate change, including melting glaciers, sea-level rise, longer agricultural seasons, changes in biodiversity, and the effects on health due to heatwaves.
- The impact of future climate change may adversely affect almost all European regions with a rigid social policy system (unfavourable social climate) and fluctuations in healthcare and infrastructure. Many sectors of the economy, such as agriculture and energy, may face difficulties.
- In Southern Europe, higher temperatures and droughts can reduce water availability, hydropower potential, summer tourism, and crop productivity, which hinder economic activity in many regions of Europe.

- In Central and Eastern Europe, summer precipitation is expected to decrease, resulting in increased water scarcity. Forest productivity is expected to decline, whereas the incidence of peat fires is predicted to increase.
- In Northern Europe, different impacts are anticipated due to climate change, including some benefits, such as reduced heating demand, increased crop yields, and increased forest growth. However, as climate change continues, the negative effects may outweigh the benefits. These would include more frequent winter floods, endangered ecosystems, and increasing soil instability.

Central and South America

- By mid-century, increasing temperatures and decreasing soil moisture are predicted to cause irreversible changes in savannah and tropical forests in the Eastern Amazon.
- In arid areas, climate change is likely to aggravate the situation, leading to salinisation of agricultural land (increase in salt content) and accelerating desertification (land degradation). In some areas, the productivity of livestock and some important crops, such as maize and coffee, is projected to decline, which will have negative consequences for food security.
- It is expected that sea-level rise will increase the risk of floods, transfer of people, salinisation of drinking water resources, and coastal erosion in low-lying areas. This threatens fish stocks, recreation, and tourism.
- Changes in precipitation and melting glaciers are projected to have a major impact on water availability for human consumption, agriculture, and energy production.
- It is believed that climate change and change in land usage will increase the extinction of species.

North America

- The warming climate in the western mountains will reduce the amount of snow, increase the number of winter floods, and reduce the amount of precipitation during the summer, which will cause the depletion of water resources.
- Pests, diseases, and fires are expected to increasingly affect forests. The level of fire load in forests will increase.
- It is predicted that at the beginning of the century, climate change will increase the productivity of rain-fed agriculture in the northern areas, on average, but by 2020, rising temperatures during the warm season will pose significant challenges for water use.

- The number, intensity, and duration of heatwaves over the century are projected to further change cities that are already exposed to heatwaves, which could have negative effects on health and increase tensions in energy systems. The population of the elderly is at the greatest risk.
- Climate change is likely to increasingly affect coastal communities and habitats, exacerbating existing problems in terms of population, development and pollution, infrastructure, human health, and the ecosystem.

Polar Regions

- Climate change is likely to reduce the thickness and scale of glaciers and ice sheets.
- Changes in natural ecosystems are likely to have adverse effects on many organisms, including migratory birds, mammals, and bigger predators, as marine species change.
- In the Arctic region, climate change is likely to reduce the extent of sea ice and permafrost, which can have unequal impacts on human settlements. Adverse effects may include damage to infrastructure and changes in winter activities, such as ice fishing and ice road transport. The positive effects may include more navigable routes in the northern seas.
- Rising sea levels can exacerbate coastal erosion and disrupt natural and social systems.
- Terrestrial and marine ecosystems and habitats are expected to be at risk from invasive species.

Small Islands

- Small islands in the tropics or higher latitudes are highly vulnerable to extreme weather conditions, sea-level changes, rising air and surface temperatures, and changing rainfall.
- Deterioration of coastal conditions, such as beach erosion and coral change, is likely to affect local resources, for example, fisheries, as well as the value for tourist destinations.
- Sea-level rise is expected to increase the risk of floods, storm surges, erosion, and other coastal hazards. This impact would threaten vital infrastructure, settlements, and facilities that support the survival of island communities.

As it can be seen from the analysis, the biggest challenges are for those regions where most of the developing economies are located.

Climate change is a classic example of market failure, or in other words, market distortion. The costs of CO_2 emissions are covered by society as a whole, and those who burn fossil fuels benefit. To tackle a market mistake, for example, by applying a carbon tax, we need to know the social costs of these CO_2 emissions. In addition, when governments assess the costs and benefits of implementing policies or investments, they need a value for CO_2 emissions. If the social costs of carbon USAGE are high, the benefits of CO_2 emissions are high, and climate mitigation actions will be costly.

If the world operates as a great economic model, the "optimal" amount of climate change mitigation efforts is reached when the expenses of the additional costs related to further emissions reductions are balanced due to limiting further warming. Again, if we are not sure about the optimal level of mitigation, it does not mean that the correct answer is "zero".

The social cost of carbon (SCC) is a key concept in understanding and implementing climate change policy. This term refers to the economic costs emerging due to additional tonnes of carbon dioxide or its equivalent. The SCC has become a key tool in climate change policy, particularly in setting regulatory policies that cover greenhouse gas emissions (Nordhaus 2014). Estimates of social costs are necessarily complex because they cover the full impact of emissions over the carbon cycle and climate change, including the economic damage caused by climate change. There are currently a number of established integrated assessment models (IAMs) that are available to assess the entire pathway of cause and effect and can, therefore, calculate a consistently synchronised value for social carbon costs.

These integrated assessment models consist of four elements:

- Firstly, there are socio-economic projections: How many people will there be in 2150? How fast will the economy grow in the next century? How much CO_2 will be emitted?
- Secondly, there is the "climate module": How does climate change respond to CO_2 emissions? How fast does the sea level or temperature rise? What to do with rainfall, what will it become like, and what to do with extreme weather events?
- The third element is the benefits and damages: How will climate change affect crop yields? What are the costs of living or adapting to changing sea levels? How do elevated temperatures affect productivity or energy use for heating and cooling? How can we assess non-market impacts, for example, loss of biological species and habitats?
- Finally, the fourth element in discounting is used to estimate future benefits and costs in today's money. Future losses tend to dominate estimates of social costs, as CO_2 remains in the atmosphere for thousands of years and damage increases as temperature rises.

The DICE model (the Dynamic Integrated Climate-Economy model) is one of the main integrated assessment models used by researchers and governments to estimate social carbon costs (Nordhaus 2014).

Figure 5.4 presents a chart showing world regions which currently suffer the most from the effects of climate change.

Figure 5.5 provides a chart showing what caused climate change historically.

Historically, with the rapid growth of national economies, increasing environmental pollution could not have been avoided, as growing industrial scale has also increased greenhouse gas emissions.

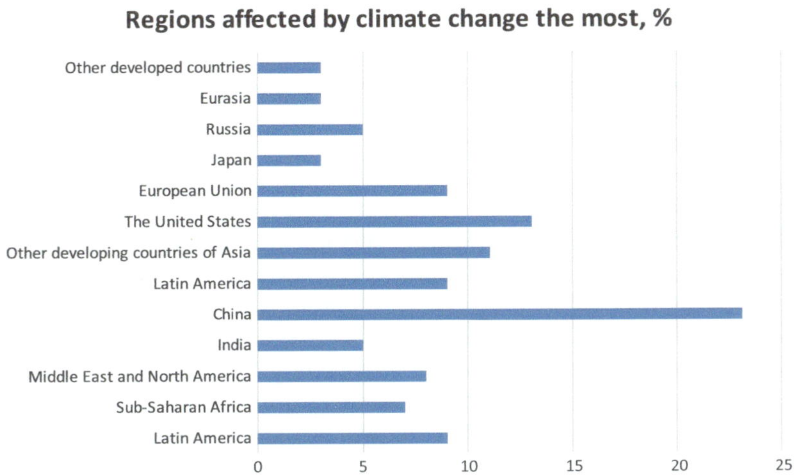

FIGURE 5.4
Regions affected by climate change the most.

Source: National Research Council 2016.

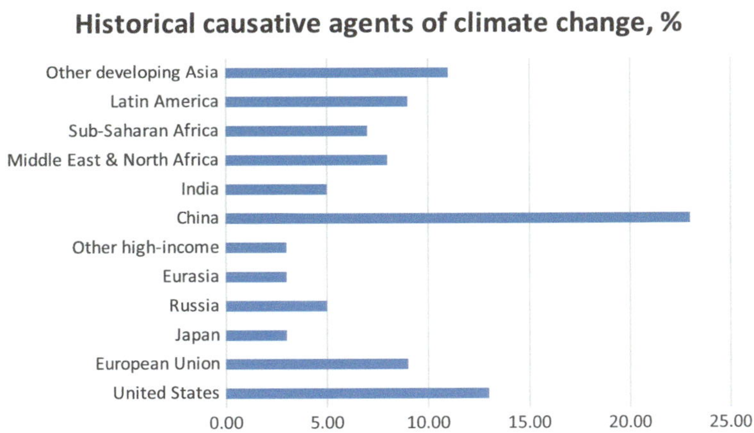

FIGURE 5.5
Historical causative agents of climate change.

Source: CO_2 emissions excluding LUCF 1850–2011, (CAIT v.2.0) National Research Council 2016.

Figure 5.6 provides a diagram showing which countries are currently responsible for climate change and to what extent.

Currently, more than 1 billion people with two-thirds of women live in extreme poverty, i.e. less than one dollar per day. The vast majority of them are in developing economies.

While global communities were unable to prevent climate change and its effects, developed and developing countries could have worked together to mitigate climate change. Developed countries should provide financial subsidies to address the effects of climate change in developing countries, especially in areas that are directly related to survival (e.g., the agricultural sector). Developed countries can provide human resources to developing countries to train local communities and share best practices on how to combat the effects of climate change and how to adapt to the already existing effects of the process. Climatologists from developed countries can successfully join such activities by providing public education services to the population and policymakers of economically less developed countries. Another, no less important form of aid would be an investment in research, which should be carried out more rapidly now than ever in the regions most affected by climate change. The results of the research would help to correctly assess the current situation and predict prospective development plans, taking into account the effects of climate change and other possible changes. Research reports would be a key recommendation for all interested parties, including communities in developing countries, which need to develop the capacity to adapt to climate change. Only by focusing on joining efforts, material, financial, and human resources, a further perspective of the developing world could be foreseen.

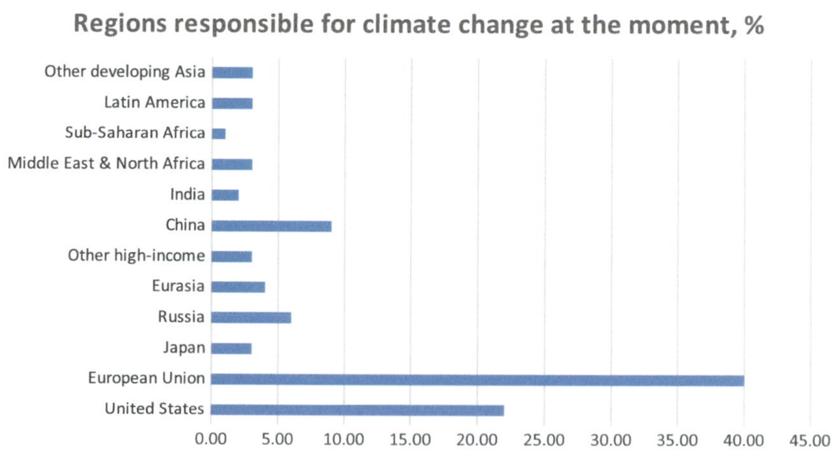

FIGURE 5.6
Regions responsible for climate change at the moment.

Source: CO_2 emissions including LUCF 1850–2011, (CAIT v.2.0) National Research Council 2016.

5.3 Climate Change and Cities

The urban population is steadily increasing in absolute terms while the rural population is declining in all parts of the world. According to the United Nations Population Fund (2007), the world is experiencing the largest wave of urban growth in history. There were only two cities in the world with a population of more than a million in 1800. By 1950, there were already 75 cities of this size, and by the year 2000, there were already 380 cities with more than 1 million people, half of which were in Asia (Huq et al. 2007). In 2000, the average population of the 100 largest cities was 6.3 million. In 2008, more than half of the world's population lived in or around cities. By 2030, this number is projected to reach nearly 5 billion, with urban growth concentrated in Africa and Asia (Bicknell et al. 2009).

Every year, already existing cities increase in population by 67 million new urban residents. Much of this rapid urbanisation is taking place in developing countries (see Figure 5.7). By 2030, about 60% of the world's population will live in cities. This figure will remain until the middle of this century when it is projected that more than 65% of the world's population will be living in cities (IPCC 2014a–g). This rapid, unprecedented urban growth poses many challenges for cities, even without additional solutions to climate change. However, while the world's urban population is growing, cities, and the people living, there are becoming increasingly vulnerable to extreme weather events, including more frequent, longer, and more intense heatwaves, and inland floods due to heavy rainfall and coastal floods caused by sea-level rise.

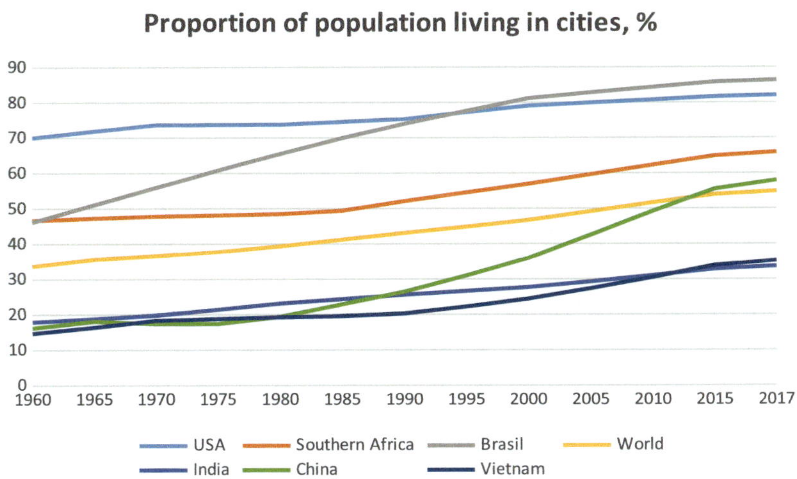

FIGURE 5.7
Proportion of population living in cities.

Source: World Bank.

Climate change is already happening, and cities increasingly need to plan and implement solutions to the various climate risks that can affect the health and well-being of the population, as well as the city's economy and infrastructure systems. 70% of cities are already facing the effects of climate change, and those effects are risky. For instance, more than 90% of all urban areas are on the coasts, making many cities around the world at risk of being flooded by sea-level rise and powerful storms. The financial effects of climate change can be as devastating as the physical and unexpected consequences of storms, floods, and droughts.

The chart below (Figure 5.8) provides an overview of how urbanised the world has become and what prospects for urbanisation are in the future. The figure shows that more and more people are living in cities.

As can be seen from Figure 5.8, the urban population exceeded the rural population around 1920. Such tendencies have persisted to this day. Moreover, the share of the urban population is projected to grow in the future, with 89.16% of the world's population living in cities by 2050 (see Figure 5.9), representing around 6.68 billion people (see Figure 5.10).

With more than half of the world's population living in cities, they need to address issues caused by climate change.

Cities are particularly vulnerable because they are "moving". Such urban infrastructures as bridges, metro systems, buildings and roads, a historic sense of the place, and the establishment of the population are very important elements of cities. Nevertheless, the strengths of such an area can become a

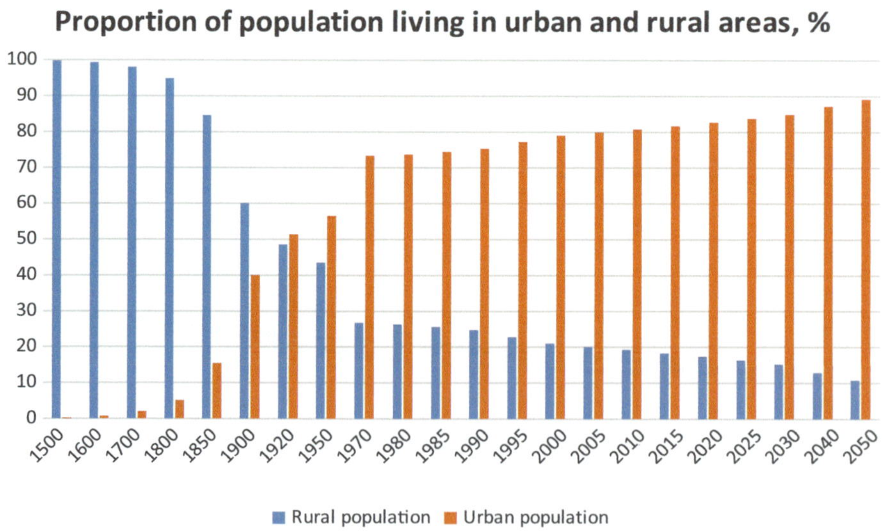

FIGURE 5.8
Proportion of population living in urban and rural areas.

Source: Our World in Data https://ourworldindata.org/urbanization.

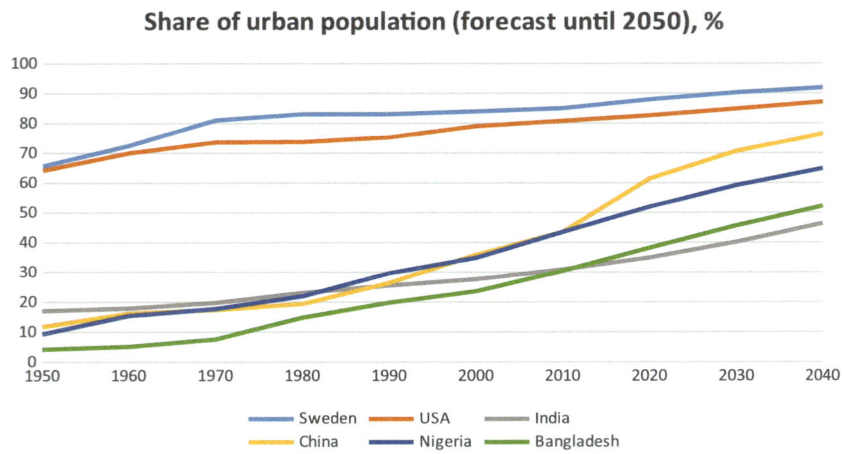

FIGURE 5.9
Share of urban population (forecast until 2050).

Source: UN World Urbanization Prospects 2018.

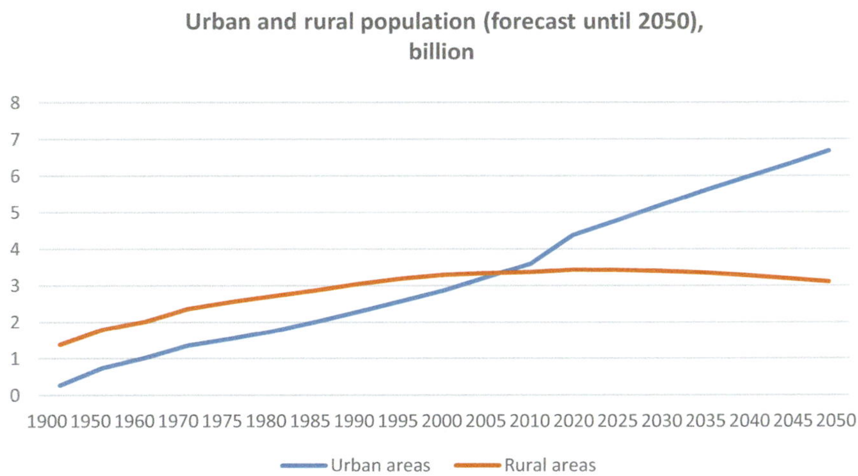

FIGURE 5.10
Urban and rural population (forecast until 2050).

Source: UN World Urbanization Prospects 2018.)

commitment if the local ecosystems on which they rely are unable to adapt to climate change. Climate change poses a serious threat to urban infrastructure, quality of life, and all existing urban systems. Not only poor but also rich countries will be increasingly affected by abnormal climate phenomena and climate change tendencies (World Bank 2010).

Urban areas are severely affected by climate change and major factors of greenhouse gas emissions. Cities play a key role in adapting to climate change.

Urban areas emit up to 70% of the world's anthropogenic greenhouse gases into the atmosphere, with fossil fuel combustion being the main source. In order to achieve the 2015 Paris Agreement on climate change under the objective of maintaining the temperature significantly lower than 2°C compared with pre-industrial period, it is necessary to keep the "carbon budget" and emit into the atmosphere no more than about 800 gigatons of CO_2. An important step in analysing urban greenhouse gas emissions is the development and use of standardised methodologies. Such methodologies allow the comparison of the cities and provide information on the cities' contribution to national emissions. Officials in the cities are increasingly taking action by setting city-level emission reduction targets or using other measures, such as expanding the use of renewable energy technologies.

Leaders of the cities are in the right position to address climate challenges for two main reasons: first, cities are major energy consumers, consuming two-thirds of the world's energy while generating more than 70% of greenhouse gases. Second, city mayors in many countries are directly accountable to their constituents for their decisions that affect global change and daily life.

Table 5.2 summarises the results of technical analysis report conducted by the Urban Climate Change Research Network (UCCRN) of 2018.

The technical analysis carried out by UCCRN includes three main components:

1. the analysis of climate damage data from six major cities on a global scale;
2. a case study that contextualises the six key vulnerabilities and how cities respond to them; and
3. the geographic information system (GIS) database that assesses the layers of maps that can be used to assess these six hazards in different cities around the world.

This technical report includes the results of a global data analysis, global city maps that are expected to be exposed to vulnerabilities, and case studies based on research derived from tangible urban experiences.

5.3.1 Extreme Heat

As shown in Table 5.2, in 2018, approximately 350 cities in the world with an average 3-month temperature of at least 35°C experience extreme heat conditions; about 200 million people in cities live in extreme heat, making up 14% of the world's population. In 2050, more than 970 cities are projected to be regularly exposed to the hottest 3-month average temperatures of at least 35°C; more than 1.6 billion people in cities will live in particularly high

TABLE 5.2

Estimated Vulnerability of Cities and Their Inhabitants

Vulnerability	Period	Number of Vulnerable Populations	Number of Vulnerable Cities
Extreme heat: common for people living in cities where they are regularly exposed to the hottest 3-month average maximum temperatures that are present today and will be in 2050.	Present	More than 200 million people	More than 350 cities
	2050	More than 1.6 billion people	More than 970 cities
Extreme heat and poverty: the total number of people living in poverty in cities where they are constantly exposed to the hottest 3-month average maximum temperatures that are present today and will be in 2050.	Present	More than 26 million people	More than 230 cities
	2050	Nearly 215 million people	More than 490 cities
Water availability: the total number of people living in cities where water flow (availability) will decrease by at least 10% by 2050 compared with the current day.	2050	More than 650 million people	More than 500 cities
Food security: the total number of people living in cities where the national yield from at least one of the four main crops (maize, rice, soybeans, or wheat) is projected to fall by at least 10 % by 2050 compared with today.	2050	More than 2.5 billion people	More than 1,600 cities
Sea level rise: the number of people living in coastal cities with a projected water rise level of at least 0.5 m by 2050 compared with the current day. Coastal cities are defined as being within 10 km from the coast.	2050	More than 800 million people	More than 570 cities
Sea level rise and energy supply: the number of people living in cities that are close to energy supply facilities located 50 km from the city by 2050.	2050	More than 450 million people	More than 230 cities

Source: UCCRN Technical Report (2018).

summer temperatures, and this will make up 45% of the world's urban population. The number of people living in cities that are constantly exposed to extreme heat conditions will increase by 700% compared with 2018.

As global average temperature increases in the coming decades, the frequency, intensity, and duration of extremely high temperatures are also expected to increase. Cities are particularly affected by the urban heat island effect (UHI). The UHI effect is a phenomenon when the average temperature in a particular region is higher than in the surrounding areas. This is most often the case when comparing the climate of cities with nearby rural areas. In such cases, the temperature differences are mainly due to the fact that the materials used in the construction of buildings and roads absorb and retain more solar heat than in nature. In addition, more heat is emitted in cities due to human activities and buildings trap wind that can disperse it (Gartland 2011). As a result, urbanistic centres are more responsive to extreme events that can worsen air quality and lead to population dehydration, heat strokes, cardiovascular complications, kidney diseases, and death (Knowlton et al. 2014). Certain groups of people, including children, the elderly, the sick, and the poor, are particularly vulnerable to constant heatwaves.

By 2050, people who already live in hotspots will have to adapt to even longer hot periods; meanwhile, people living in cooler cities will face extremely high heat levels that are unusual for them. Consequently, cities will need to adapt to rising temperatures to help residents move easily and safely outdoors as well as find solutions for air conditioning and ventilation. Ozone production increases at high temperatures, and short-term exposure to ozone is associated with adverse health effects. High ozone concentrations are also associated with chronic lung diseases and increased mortality rate (Bell et al. 2004). Children, the elderly, the sick, and the poor living in cities are particularly vulnerable to heatwaves (Wilhelmi et al. 2004).

The direct effects of heat are especially harmful when there are high temperatures at night. A person cannot rest properly under such conditions as an increase in body temperature under heatwave conditions greatly reduces the heat loss required to adjust the internal mechanisms that regulate body temperature. Such a condition can lead to body fatigue (Amengual et al. 2014). Indirect effects of heat on health result from the interaction of heat and other environmental factors, especially, air and water pollution. For example, air pollutants and heat can cause higher ozone concentrations, which can irritate the respiratory tract and reduce lung function. As a result, heatwaves can cause heart attacks, asthma, bronchitis, and other cardiopulmonary diseases that can be fatal (Franchini and Mannucci 2015).

Given the complex evolution of heat extremes and the synergy of the abovementioned physiological factors, a variety of causes and effects will need to be taken into account for each city's heat-related action strategy in order to strengthen the resilience of cities to future climate change. In a warming world, it is vital for adaptation strategies to include more efforts to improve urban infrastructure and services. A resilient city must have access to affordable healthcare, reliable public transport, uninterrupted electricity supply, clean drinking water, and well-functioning sanitation systems (including air conditioning and ventilation). By reducing emissions of pollutants, cities around the world can work together to ensure that the vision for the future is not pessimistic.

5.3.2 Extreme Heat and Poverty

More than 26 million people live in poverty in more than 230 cities, where they are constantly exposed to the hottest 3-month average maximum temperatures reaching at least 35°C (see Table 5.2).

By 2050, nearly 215 million people will live in poverty in 495 cities, where they will regularly face the hottest 3-month average maximum temperatures of at least 35°C (Table 5.2).

The number of people living in poverty will increase eightfold compared with conditions of today. The effects of climate change on people and groups vary due to different socio-economic factors. People with low incomes and limited savings, people who are discriminated based on their race, ethnicity, gender, age, and poor health, often experience poverty and social exclusion, making it the hardest for them to survive climate change. The impoverished population is particularly vulnerable to climate change due to a lack of access to resources that can help them cope with extreme situations. For example, people living in poverty during the extended heat period face additional risks, compared with others, because they may lack drinking water, shelter, air conditioning, and medical care. As the urban population continues to grow, the number of people vulnerable to climate change grows too.

Nevertheless, the effects of climate change will have some impact not only for the most vulnerable. Regular climate change events can also gradually affect more resistant groups in society, ultimately leading to increasing levels of urban poverty (Tyler and Moench 2012; Tompkins et al. 2013). Climate change will reduce the financial resources of many people who are not currently considered particularly vulnerable. If cities face extreme climate change more often, they simply do not have time to recover, and this increases the overall level of poverty in the city (Tyler and Moench 2012).

Differences in the scale and nature of the risk are related to the scale to which infrastructure elements and services are provided in a given area. Differences in risk due to inadequate or incomplete infrastructure and lack of access to services can be due to factors such as age, gender, and health status. For instance, the 1991 Bangladesh cyclone killed 138 thousand people, many of them were women over the age of 40 (Dankelman et al. 2008). Also, the analysis of the effects of the floods in 2011 in Lagos Island revealed the vulnerability of low-income women, in terms of gender relations and gender roles in household structure, occupation, and access to healthcare (Ajibade et al. 2013). Differences in risk may also be due to the inactivity of certain groups, such as informal settlements, and the lack of accountable government agencies in certain areas (Huq et al. 2007).

Growth rates are particularly rapid in many cities affected by higher-than-average impacts of climate change. These cities are mainly located in poor countries in Asia and Africa. It is estimated that by 2050, about 90% of the country's population will live in these cities. These demographic trends are worrying, as the growing urban population will also increase the number of people affected by climate change and raise issues of justice and, in particular, environmental justice that will need to be addressed. Rapid population growth in cities in developing countries is likely to correspond with the increasing slum populations, which are often associated with unplanned and unregulated settlements in risk areas (UN-Habitat 2013; Revi et al. 2014).

To mobilise resources in order to improve environmental justice in the circumstances of a changing climate requires the active participation of climate-affected communities and the involvement of civil society in decision-making. The key is to learn to take advantage of non-traditional sources of funding, including partnerships with the private sector and potential community input. Governance must be also strengthened in terms of costs and compliance with the principles of transparency in monitoring and evaluation (Newell 2005).

Working with environmental groups can make a significant contribution in raising public awareness, emphasising equality issues and promoting the inclusion of justice in urban plans and policies. Special attention should be paid to the implementation of the principles of equality between women and men. As noted in the United Nations report, "the urban poor have a proven capacity to improve and invest in their communities". Cities engaged in an open dialogue with vulnerable communities can often find innovative and sustainable ways to build resilience to climate change (Rosenzweig et al. 2018). For climate policy to be effective in the long term, it is essential to ensure that development policies and climate change adaptation plans are tailored to the poor and society as a whole.

5.3.3 Water Availability

Due to climate change, water is both a resource and a threat. Good quality water is a key source that ensures the well-being of a growing urban population. However, droughts or excessive rainfall can lead to risks such as damage to physical property, reduced water quality (adverse consequences for health), and lack of proper water flow to the sewers. As cities grow, demand and competition for limited clean water resources will increase. Climate change in cities around the world will only exacerbate the problems associated with clean water (Vicuña et al. 2018).

Freshwater is globally considered strategic because it is an extremely important element in the development and security of every country. Freshwater not only ensures the maintenance of life but is necessary for social and economic activities, which manifest through energy and food consumption, the development of individual industries, and waste disposal. Freshwater can be considered a strategic resource also because in many countries there is a direct dependence between the development of the state and the development of the agricultural sector or on hydropower.

Climate pressure interacts at different spatial scales and has a synergistic effect on water availability, which depends not only on water quantity in different sources but also on water quality, infrastructure integrity, competing consumer agreements, and the strength of institutions. Management systems failed to adequately address the challenges that climate change poses to urban water security. Failure is often due to a lack of coherent and responsible policies, limited technical capacity to plan adaptation, limited resources to invest in projects, a lack of coordination, and a low level of political will and public interest (Vicuña et al. 2018).

Higher temperatures result in higher water demand in many cities (Schleich and Hillenbrand 2009). However, the extent of this temperature sensitivity is highly dependent on climate, land use, and energy dependence in cities (Zhou et al. 2000; Ruth et al. 2007; Donkor et al. 2014; Stoker and Rothfeder 2014). For example, climate change may have a greater impact in cities that depend on older, less water-efficient, and carbon-based thermal power plants than in cities that rely on newer, more water-efficient, natural gas-powered combined cycle thermal power plants (Scanlon et al. 2013a, b).

Cities often obtain water from sources far away from their local water supply systems, making urban water supply highly dependent on climate change in the surrounding areas. Different water uses and urban users have different requirements for water supply, wastewater, and sanitation. In the city centres, key services, such as healthcare, food supply, transport, energy systems, schools, and retail, are interrelated with water. The projected future deficit of urban water supply could have a significant impact on both water availability and costs. The decisions made now will have a major impact on the future supply of water to industry, households, and agriculture.

Strategies for adapting urban water resources are unique to each city, as they are highly dependent on local conditions. Therefore, understanding the local context is essential for adaptation of water systems in such a way that current and future climate risks are taken into account. Urban water insecurity, especially in low-income countries, is a constant challenge, and water scarcity is also often a catalyst for conflict. Many cities put the effort in providing basic services to their residents, especially those living in informal settlements (which are especially relevant for cities in developing countries). As cities grow, demand and competition for limited water resources will increase, and climate change might exacerbate the situation. Therefore, action taken now can reduce adverse effects in the long run. Principal planning of water supply systems should take into account projected climate change over a period of at least 50 years. However, given the uncertain future, finance and investment should focus on opportunities that promote water security and economic development, and policies should be flexible and responsive to changes and new information that emerge over time.

Water availability is often the first victim due to the effects caused by climate change. If urban authorities do not act quickly to meet their mitigation and adaptation obligations under the Paris Agreement, water scarcity, which affects more than half a billion people living in 500 cities, would cause socio-economic problems, damage the urban economy, and threaten their stability.

5.3.4 Food Security

There are more than half of the world's population living in the cities, but the production of food is very little. As a result, cities need complex systems to feed their growing populations. While much of the world's food is supplied in commodities, much of the food consumed by the urban population comes from national sources. As climate change is projected to reduce crop yields in many parts of the world, these reductions will affect the food security of many urban residents (Rosenzweig et al. 2018).

The effects of climate change are a concern for agricultural production and its impact on the urban population, as changes in the quantity (which is steadily declining) and quality of agricultural production pose a threat to food security. Declining crop yields will reduce the availability of food in cities and affect its availability as prices rise. Thus, the welfare of urban citizens can be affected in a variety of ways.

With increasing urbanisation, cities face the threefold challenge of mitigating climate change, adapting to climate change and providing basic goods and services to the population, including food. The world is already experiencing the effects of climate change, with the global average temperature increasing by 1 degree Celsius more than before the Industrial Revolution. The results are greater differences in rainfall, droughts, and other extreme events that have affected agricultural production and quality. If crop yields fall and prices rise, it will be a major challenge for residents and local authorities around the world. By 2050, the city's population, especially in developing countries, is expected to grow. Therefore, estimates suggest that agricultural output should increase by 50% to 60%, respectively (Reardon 2016).

In a world characterised by rising temperatures, droughts, rainfall swings, and other extreme climatic events, food security is a matter of health, public security, and social justice. As the impact on the climate becomes more and more apparent, urban food supply systems need to become more resilient, and urban governments need to play a major role in ensuring food security for the population.

5.3.5 Sea-Level Rise

Sea-level rise, coupled with rapid urban development, has significantly increased the risk in coastal cities. Coastal regions are vulnerable to climate change, such as sea-level rise, storms, coastal erosion, and saltwater infiltration. 10% of the world's population and 13% of the urban population live in low-lying coastal areas defined as adjacent coastal areas that are 10 m above sea level (McGranahan et al. 2007). Most mega-sites are located along the coasts, which are vital for shipping, fisheries, and international trade. Rising sea levels are threatening coastal populations, the infrastructure of particular importance, and valuable coastal assets.

Major coastal cities often have valuable assets that are located in the coastal zone, including port facilities, transport and utility infrastructure, schools, hospitals, and other long-term structures. These assets might be at risk of both short-term floods and permanent flooding. It is estimated that global economic costs allocated to cities due to rising sea levels and inland water floods can be as high as 1 trillion dollars per year (Hallegatte et al. 2013). By 2070, the impact of climate change on assets due to global sea-level rise, land rise, population growth, and accelerating urbanisation could increase more than tenfold and cover about 9% of projected global GDP.

Urban centres built in the valleys are particularly vulnerable. They are at immediate risk of flooding, which is linked to the threat to lives and property. This threat will be exacerbated by rising sea levels, which will allow storms and other unpredictable natural phenomena to spread deep into the continent and reach inland waters. Moreover, this risk increases in many coastal cities due to changes in the subsoil caused by overuse of groundwater and the formation of sediments.

In addition, the salinity intrusion into the upstream and coastal aquifers caused by sea-level rise could jeopardise drinking water supplies in the cities and contaminate agricultural soils. Sea-level rise may also increase the potential of groundwater. High waves and/or water levels during intense storms can cause beach erosion and retreating shores.

A long-term integrated approach to coastal management and strategic urban management is essential for adapting to climate change and managing coastal zone cities. Actions to reduce natural hazards include the displacement of people and infrastructure from hazardous areas to "hard" engineering protection zones and "soft" solutions, such as planting vegetation in coastal areas (Möller et al. 2014).

5.3.6 Sea-Level Rise and Energy Supply

Urban energy systems are directly and indirectly vulnerable to climate change, including the effects of heatwaves, droughts, and floods. Rising sea levels pose a particularly serious threat to electricity supply, as infrastructure systems designed to supply electricity often operate for decades and are often physically outdated. As sea levels continue to rise, the effects of floods will be increasingly experienced in coastal regions. Many cities depend on energy produced by power plants located in low-lying coastal areas. Floods can have a cascading effect on electricity generation, transmission, and distribution networks, which is reflected in the urban population and key infrastructure systems. It is difficult and very expensive to move already built and operated electricity networks to other, safer places, and in some cases, it is simply not possible. Understanding the risk, which will be posed to electricity networks by sea-level rise, the potential impact of floods must be taken into account before new energy infrastructure elements are deployed.

There are currently 270 power plants near the cities, generating a total of 182,902 MW of electricity, which will be directly vulnerable if sea levels rise by at least 0.5 m by 2050. These power plants supply electricity and heat to more than 450 million people and more than 230 cities (UCCRN Technical Report 2018).

Energy supply is an essential aspect of global economic and social progress and an essential element of urban security and resilience. Cities consume more than 75% of the world's primary energy, and the biggest losses are due to climate-related energy disruptions (UN Habitat 2013). Such disruptions in energy supply in cities can affect the provision of electricity, transport, water, healthcare, and other important services and cause negative gradual changes throughout the city and country economy.

Due to the projected rapid population growth, urbanisation, and climate change in the coming decades, it is necessary to strengthen already existing and planned energy system infrastructure. Looking ahead, when choosing new energy production sites, local air quality, greenhouse gas emissions, the correct way of energy supply, energy security, and the system's resilience to climate change should be taken into account. Municipalities will have to carry out vulnerability assessments and plan for various climate events and their impact on the energy supply chain. This is particularly relevant for coastal energy infrastructure, as it is most vulnerable to climate change. More frequent extreme weather conditions, such as hurricanes, storm waves, and coastal floods, are expected to cause recurrent energy grid disruptions, reduce the reliability of production and transmission, and increase energy costs in cities, which will directly affect consumers (Azevedo de Almeida and Mostafavi 2016).

Coastal areas are also vulnerable to oil and gas drilling and oil refining operations.

5.4 The Impact of Climate Change on Agriculture

Climate and weather conditions have been reflected in people's lives since ancient times. Meetings usually begin with a discussion of the weather, what it is today and what it will be in the future, and the media often makes references to weather and climate conditions many times a day. Weather reports haunt us constantly. In addition to these common social interests, the climate has also been important for the scientific community for centuries. Scientists have long proved not only that the climate is the basis of human civilisation but also that it affects its quality of life, success, and failures (Glacken 1967; Fleming 1998).

Climate change and its changes concern humans. Recurrent droughts and floods pose a serious threat to the survival of billions of people, especially those who depend on the results of the activity of the agricultural sector. Due to extreme events such as droughts and floods, heatwaves and cold spells, forest fires, and others, the chemical composition of the atmosphere also changes.

Global warming not only causes a change in average temperature and rainfall but also increases the intensity of floods, droughts, heatwaves and typhoons, and hurricanes. The impact of climate change is also seen in various other forms throughout the world, including sea-level rise, shrinking of glaciers, movements of plant habitats northward, geographical changes in animal habitats, rising ocean temperatures, shortened winter, and early spring.

Since global warming affects not only ecological systems but also human life, it has become an important issue both nationally and internationally. Solutions to global warming are broadly divided into mitigation measures, focusing on reducing and absorbing greenhouse gases, reducing causal factors, and mitigating the effects of adaptation to climate change. Until now, the issue of global warming has focused on reducing greenhouse gases, based on international environmental conventions, for example, IPCC and Kyoto Protocol. However, in terms of agriculture, the main focus was placed on the adaptation to the effects of climate change, taking into account the assessment of the impacts and vulnerability of climate change. The IPCC stresses the importance for the agricultural sector to adapt to climate change. The reason for this is that even if the greenhouse gas emissions are reduced, global warming will continue for several decades due to the emitted greenhouse gases in the past.

Agriculture is the branch of the economy in which the land is used to obtain (grow) food, to produce food resources, and to work and process them. This branch of the economy is most dependent on climate and weather conditions and is therefore very sensitive to climate change. At the same time, sustainable farming systems can reduce agriculture-related greenhouse gas emissions and serve as a key tool to stabilise and change the trajectory of climate change, while continuing to provide food, feed, fibre, and energy.

The relationship between agriculture and climate is complex. While most greenhouse gases can be related to the use of fossil fuels for energy, e.g. burning coal and other fossil fuels for electricity and burning gasoline, diesel fuel in cars, agriculture also play an important role. For example, agriculture has always contributed to the direct emissions such as methane from animals and carbon dioxide from the soil; however, because of how the animals are fed, reared, and how much of land is used, it contributes to the increased amount of greenhouse gas emissions. A relatively new greenhouse gas threat is nitrogen oxide, which occurs naturally; however, in this period, due to the growing use of synthetic fertilizers, it has significantly increased in quantities. It is estimated that direct agricultural emissions amount for 13.5% of total GHG emissions.

Agriculture is our main source of food and is therefore very important for human survival; however, its importance and impact on the environment and climate have been studied little. Land used for farming, including the area of crop and pasture, covers almost 38% of the world area (Figure 5.11); agriculture consumes about 70% of water (irrigated agriculture accounts for 20% of total arable land and this area produces almost 40% of the world's total food production) and employs about 40% of the world's population. It shows that any changes in agriculture, whether caused by people or climate, resonate throughout the world and the economy.

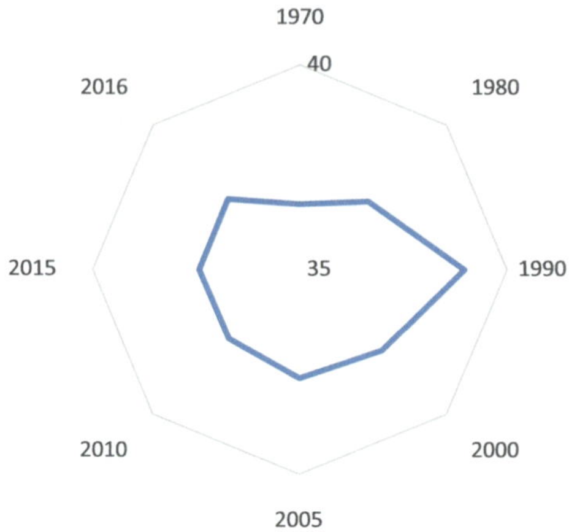

FIGURE 5.11
Agricultural land (% of the total land area).

Source: https://data.worldbank.org/indicator/ag.lnd.agri.zs.

Globally, climate variability and change may have a small overall impact on overall food production; however, regional impacts can be significant and variable, as some regions benefit from climate change and from other regions affected by climate change. In general, food production may decline in many regions of utmost importance (such as subtropical and tropical areas), and due to climate change, the agriculture in developed countries may be more productive than before depending on which countries use adaptation measures and technologies.

The diagram below (see Figure 5.12) shows agriculture value added per worker (measured in USD in 2010). The agriculture value added per worker shall be calculated as the total agriculture value added divided by the number of people employed in agriculture.

Agricultural production is obtained by the use and selection of crops suitable for the climate of very specific regions and by the application of suitable farming methods. Therefore, agriculture is like an independent biological industry with distinctive regional characteristics. Regional characteristics refer to the characteristics of the ecosystem, which are determined by the climate of that region. Climate change disrupts the ecosystem of the agriculture, resulting in changes in agricultural climate elements, e.g., temperature, precipitation, and sunlight, which further affect the quality of arable land, livestock, and hydrology sectors.

The impact of climate change on the agricultural sector is illustrated in Figure 5.13. In particular, the effects of climate change on arable land quality and the livestock sector are known due to biological changes, including changes in flowering and harvesting seasons, changes in soil quality, and the relocation of arable land to more favourable areas. Climate change affects the

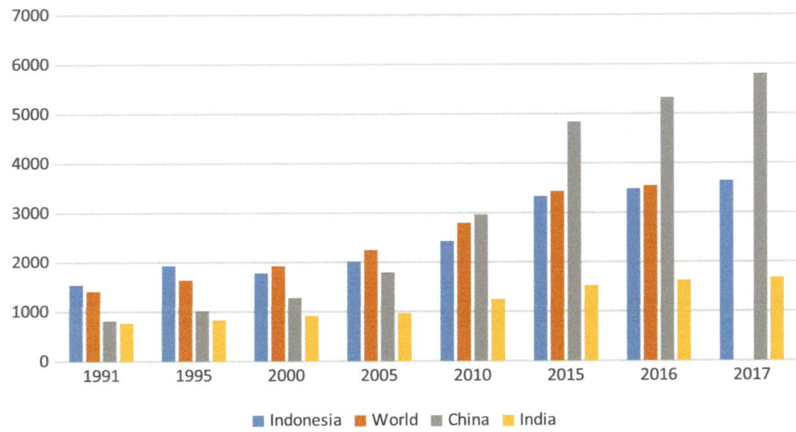

FIGURE 5.12
Agriculture value added per worker, USD.

Source: https://data.worldbank.org/indicator/ag.lnd.agri.zs.

agricultural ecosystem in such a way that pests of agricultural production can move more freely from place to place and biodiversity itself is changing. In the livestock sector, climate change is causing biological changes in areas such as fertilization and livestock breeding, as well as growing pasture area (pasture areas may be declining and this will no longer be enough to maintain the grazing area necessary for livestock farming).

Climate change also affects hydrology, including groundwater levels, water temperature, river flows, lake, and wetland water quality. Precipitation, evaporation, and soil moisture are affected. In particular, with increased rainfall due to climate change, water runoff increases, and the temperature rise increases water evaporation, which reduces water runoff. To understand the quantitative impact of climate change on water resources, a deterministic hydrology model based on the general circulation model is used.

As mentioned above, climate change has a significant impact on the rural economy, including agricultural productivity, household income, and property value, and also affects agricultural infrastructure as agricultural resources change. To date, quantitative analysis of the impact of climate change on the agricultural sector has been experimental, focused more on "cross-sectional" study. Experimental analyses are carried out on the basis of agroeconomic modelling methods, which are similar to controlled experiments in which regulated variables are similar as they are related to greenhouse gases, e.g., temperature levels and carbon dioxide emission levels. These experiments can also assess the impact of climate change on agricultural production.

The analysis of the agroecological zone is carried out using the method of crop modelling, which is often referred to as the short crop model and

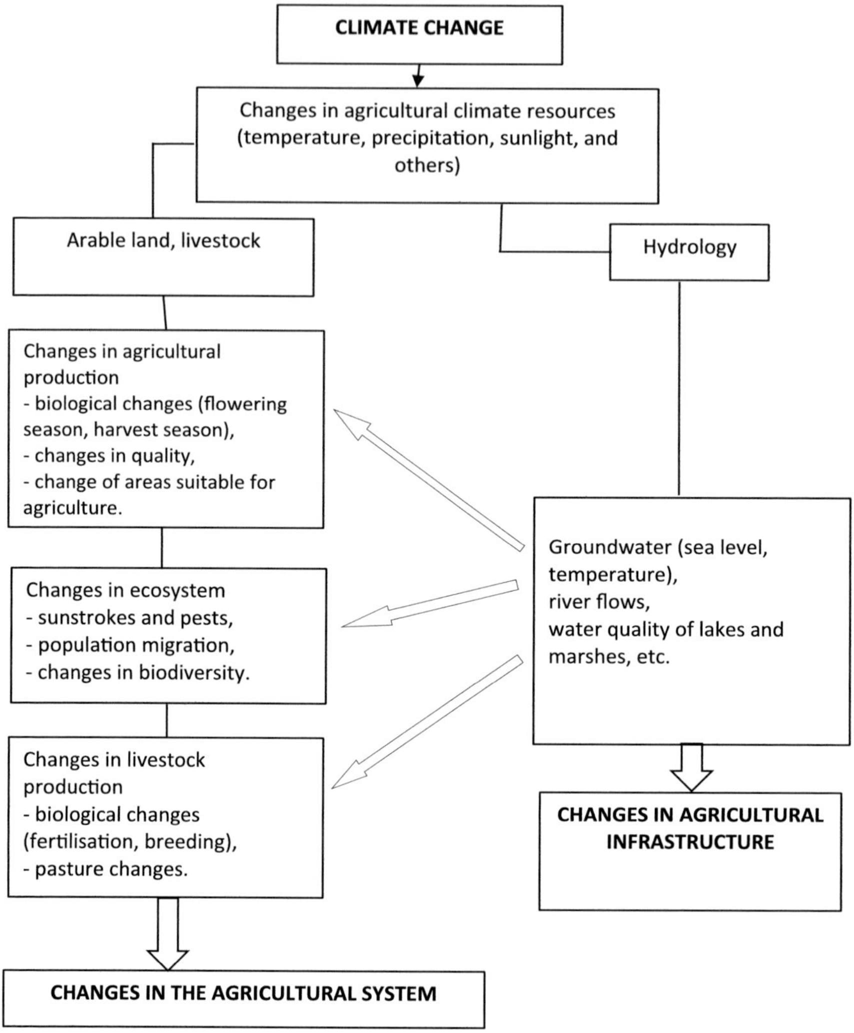

FIGURE 5.13
The impact of climate change on the agricultural sector.

Source: Created by authors.

which tracks changes in agricultural production and agroecological zones resulting from climate change. Plant growth is determined by the main three elements:

- the degree of crop production in the intended geographical area,
- cultivation technologies, and
- environment (climate, soil, and others).

The short crop model is realised using a computer programme that can estimate the growth of the crop and its quantity when these three elements are introduced. Using a crop model, it is possible to calculate and analyse agricultural production based on climate change. Estimated plant production by the model of Crop Estimation through the Resource and Environment Synthesis (CERES) model developed in the United States, the integration of the crop model, and the resource environment can be useful in adapting to climate change. The model is also useful in predicting expected situations in agriculture.

In the analysis of how and to what extent the change in temperature and precipitation after global warming affects the agricultural sector, various experiments and modelling are carried out and various research methods are used, both in laboratories and in nature. As the impact of climate change on the agricultural sector varies according to the interlinked variables, it is difficult to generalise certain analytical results. Therefore, attempts are being made to classify the effects of climate change into positive and negative ones, based on the results of studies that have been collected in the relevant areas to date (see Figure 5.14).

Global warming positive effects of agriculture include the increase in crop productivity due to fertilisation effects caused by the increase in carbon dioxide concentration in the atmosphere, the possibility of the cultivation of new varieties of plants and animals, and heating cost reductions in the northern regions by growing agricultural production in heated spaces.

The negative impact of global warming on agriculture manifests by a decrease in agricultural production due to temperature rise, spread of pests, increase in soil erosion, etc.

In addition, different climatic and environmental conditions are necessary to grow each crop. Thus, if the climate changes, for example, the temperature rises, the suitable areas for growing certain plants will shift northward. Thus, this will be a challenge for southern regions (their crop area will decrease), though at the same time, northern regions will get a new opportunity (their crop areas will increase dramatically). In this respect, the impact of climate change on the agricultural sector is ambiguous, with a negative impact on some regions and a positive impact on others. Therefore, it is important to formulate adaptation strategies that can maximise opportunities and reduce costs, which will lead to the sustainable development of agriculture.

It is known that climate change started due to the disorder of the global climate system energy balance caused by the increase in greenhouse gas and aerosol in the atmosphere and the changes in land cover and solar radiation. According to scientists, global warming is directly related to human activity, i.e. it is of anthropogenic origin (IPCC 2007).

The measures that the agricultural sector can take regarding risks and challenges of climate change are classified into

- climate change mitigation techniques that reduce the effects of climate change and reduce greenhouse gas emissions and

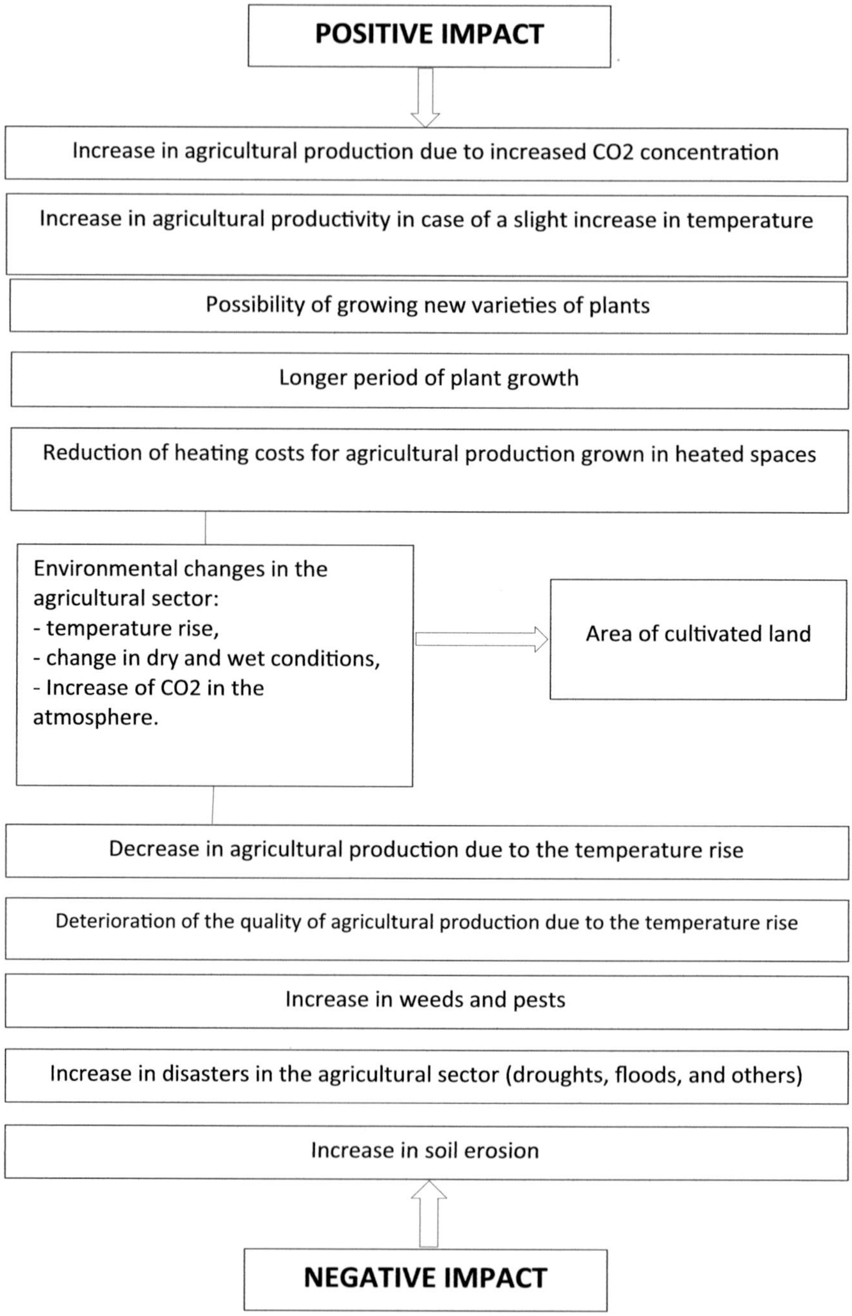

FIGURE 5.14
The potential impact of global warming on the agricultural sector.

Source: Kim 2009.

- climate change adaptation techniques when, in principle, the inevitability of global warming is recognised, the effects of climate change are assessed, and the damage that climate change can cause is mitigated (see Figure 5.15).

In the event of climate change, environmental system components (e.g., atmosphere, hydrosphere, cryosphere, biosphere, lithosphere, and others) try to adapt to changed conditions. However, if the impact of climate change is considerable, the environmental system cannot cope with the impact by self-adaptation, and additional measures are necessary. Climate change mitigation and adaptation are closely interrelated. In the long term, climate change mitigation can be seen as an adaptation tool. Adaptation to climate change is, therefore, a mandatory measure to combat climate change.

The concept of ecological (green) economic growth has been developed in order to increase policy measures in terms of sustainable development, the concept of which is an abstract and broad concept covering the

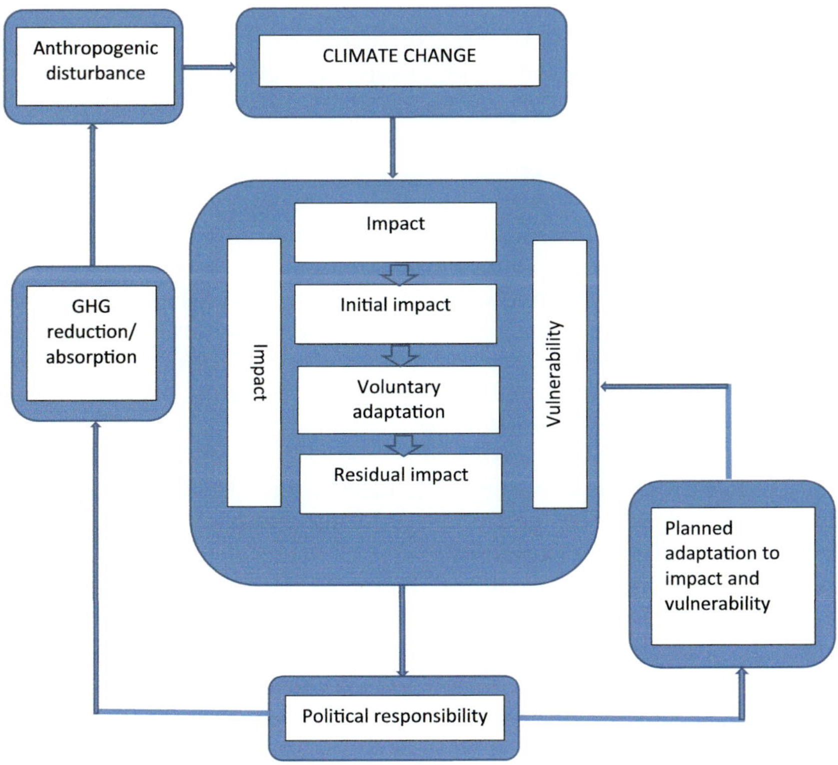

FIGURE 5.15
Approaches to measures against climate change.

Source: Created by authors.

three dimensions of economic justification, environmental protection, and social justice. The ecological (green) economic growth means qualitative growth that improves the quality of life, ensuring ecological and economic reliability.

Since there is no unambiguous definition of green growth, it can be understood as a complex and open economic growth concept, which accepts new challenges and changes in all three sustainable development dimensions: economic, environmental protection, and social perspective. In this context, ecological (green) economic growth can be seen as a concept that directly depends on the outcome of the debate about the foreseeable future.

The principle of action of green growth policy is to turn the vicious circle of the environment and economic growth into a rapidly changing cycle, thereby developing a new economic growth model and a completely new economic structure. According to the new paradigm of qualitative growth, new ideas, transformative innovations, and the latest technologies are essential factors of production. Economic growth based on these driving factors is expected to generate strong and high-quality growth, quite different from current quantitative growth.

Green economic growth constantly increases the production (including agricultural production) productivity, as "green components" (eco-technologies and knowledge) are included in the manufacturing process, thereby reducing pollution and expanding the natural capital (energy and environmental resources). For this reason, green growth leads to changes in production, technology, and consumption patterns and eco-efficiency, taking into account environmental capacity and economic and environmental aspects of production and consumption.

Green economic growth involves political decisions and assessment of the created social value, since such economic growth is achieved by the society generating low carbon emissions and green industrialisation. The realisation of green growth requires considerable costs and efforts, and most importantly, we need to change our lifestyle, moving from the nowadays consumerism to an eco-friendly and sustainable way of life. Thus, in order to realise the concept of green growth, economic incentives, development and allocation of green technologies, and understanding and cooperation of relevant state institutions are very important.

The concept of green agricultural sector development goes beyond the concept of sustainable agriculture, as it means the growth that ensures environmentally friendly and economically viable growth, which takes into account the environmental capacity of the agricultural ecosystem. The green development of the agricultural sector is achieved by moving to agricultural activities that consider the capacity of each region's environment and water system, low-carbon agriculture by reducing greenhouse gas emissions and increasing absorption capacity, and energy efficiency and savings. Green economic growth can be achieved by moving to a sustainable agricultural system, including green, low-carbon agriculture.

The transition from the traditional agricultural growth model to the low-carbon green growth model requires a very clear vision and specific goals. The appropriate vision should align agriculture with the environment, mitigate greenhouse gas emissions, and improve the agricultural ecosystem status, which contributes to the negative global warming impact reduction and improvement of quality of life for current and future generations.

In developing the green growth model, the main direction of development should be based on the 3Rs cycle reduce ↔ reuse ↔ recycle. The next step is the transition from the current maximisation of agricultural production to its optimisation. In other words, the goal of agricultural production should be shifted from increasing production to optimising production, taking into account the local agricultural environment and the conditions for greenhouse gas emissions and their absorption.

To reduce greenhouse gas emissions and improve the quality of life, a combination of support, regulation, and compensation policies should be developed, including the consolidation and coordination of environmental policies. To minimise the inconvenience and economic costs associated with green growth, green technologies should be actively developed and disseminated. Finally, it is equally important that the relevant institutions have a common understanding of the activities carried out by sharing information, education, and promotion.

The largest areas of ecologically managed agricultural land (the 2017 data) are in Oceania (35.9 million ha or 51% of global organic agricultural land), Europe (14.6 million ha or 21% of global organic agricultural land), and Latin America (8.0 million ha or 11% of global organic agricultural land) (see Figure 5.16).

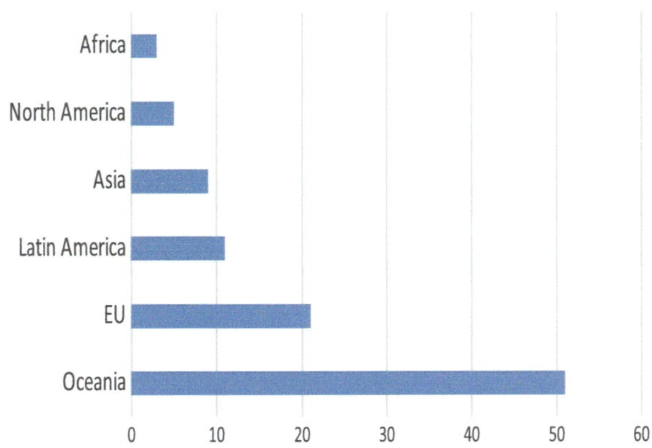

FIGURE 5.16
Largest areas of ecologically managed agricultural land, % of global organic agricultural land.

Source: https://data.worldbank.org/indicator/ag.lnd.agri.zs.

5.4.1 European Union

Countries with the largest areas of organic agricultural land are Australia (35.7 million ha), Argentina (3.4 million ha), and China (3.0 million ha). Top ten countries with the largest organic agricultural land area are shown in Figure 5.17.

In 2017, the Global Organic Food Market, according to market research company "Ecovia Intelligence" reached 97 billion US dollars (about EUR 90 billion). The United States is the leading market with EUR 40 billion, followed by Germany (EUR 10 billion), France (EUR 7.9 billion), and China (7.6 EUR billion). Top ten countries with the largest markets for organic food are shown in Figure 5.18. Compared with 2016, most major markets in 2017 grew twice.

The market for organic food in Europe continues to grow. In 2017, compared with the 2016 market, the market grew almost 11% and reached EUR 37.3 billion (European Union – EUR 34.3 million). Almost all major markets have doubled in size, and the agricultural area has increased by almost 8%. The largest ecological market in Europe is in Germany (EUR 10 billion), followed by France (EUR 7.9 billion), Italy (EUR 3.1 billion), and Switzerland (EUR 2.4 billion) (2017 data). In the world, Germany is the second-largest market after the United States (in 2017 – EUR 40 billion).

Change in organic farming area worldwide since 2000 until 2017 (million hectares) is provided in Figure 5.19.

As seen in Figure 5.19, organic farming is very popular in the world, and every year, the land area for such farming is increasing.

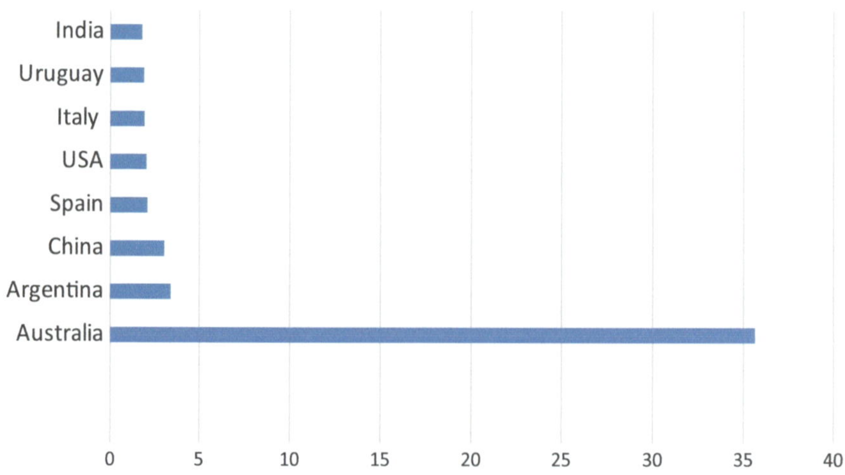

FIGURE 5.17
Countries with the largest area of organic agricultural land, million ha.

Source: https://data.worldbank.org/indicator/ag.lnd.agri.zs.

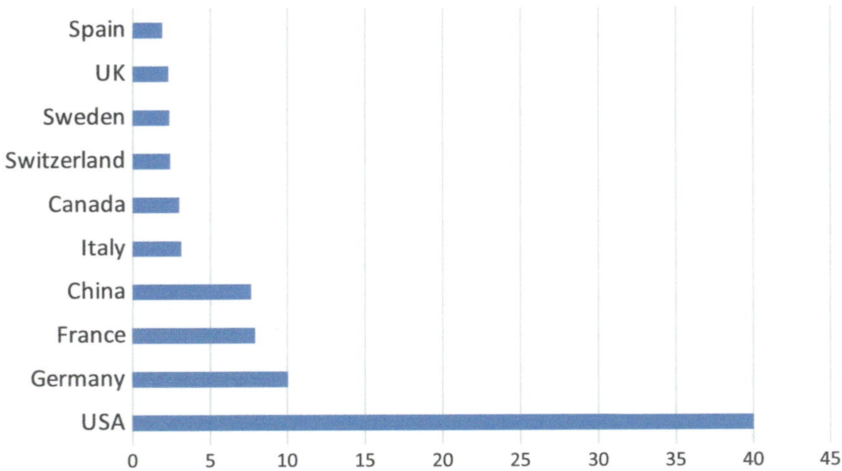

FIGURE 5.18
The largest markets for organic food in 2017, million, EUR.

Source: https://data.worldbank.org/indicator/ag.lnd.agri.zs.

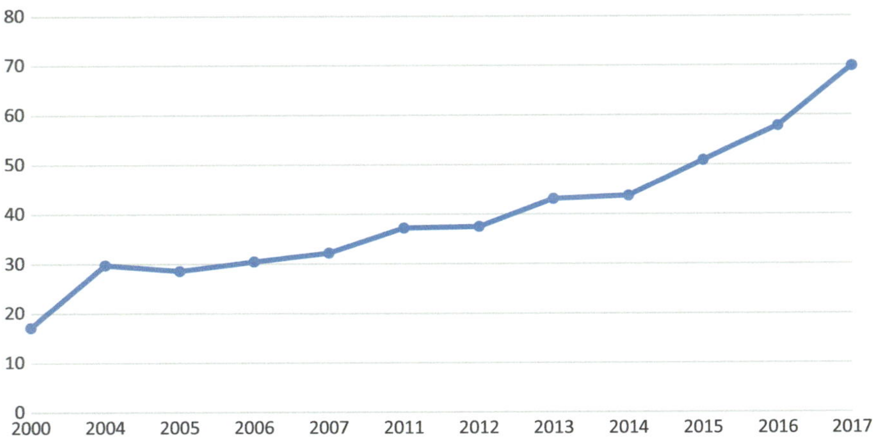

FIGURE 5.19
Organic farming area worldwide since 2000 until 2017 (million ha).

Source: https://data.worldbank.org/indicator/ag.lnd.agri.zs.

Conclusions

Climate change is one of our biggest environmental, social, and economic threats in today's world. Due to the increasing greenhouse gas emissions, the average temperature of the air and the oceans rises, snowfields and glaciers are melting faster, and the average sea level rises; as a result, natural disasters such as heatwaves, droughts, and floods are expected to recur and intensify.

The main greenhouse gas emission sources are the combustion of fossil fuels in electricity generation, transport, industry and domestic sectors, changes in agriculture and land use, landfilling, and the use of industrial fluorinated gases.

To combat climate change, countries of the world by signing the Paris Climate Agreement in 2015 committed to take action and implement climate change mitigation policies and stabilise greenhouse gas emissions, based on the best available scientific knowledge, to ensure that global temperature would increase well below 2°C, compared with pre-industrial level, and continue to keep the temperature rise of no more than 1.5°C, compared with the pre-industrial level.

Globalisation and increased international trade help the world to specialise and become more productive; however, the real efficiency is achieved when environmental damage is reduced and economic growth is decoupled from resource use. The Paris Climate Agreement not only contributes to the fight against climate change and encourages the transition to clean energy but also opens up new business opportunities for EU clean energy technology undertakings. It gives a competitive advantage to those industries that have already reduced their carbon footprint. The key task now is to complete to organise the set of implementing rules for the agreement that will ensure that all countries contribute to reducing greenhouse gas emissions.

The main climate change mitigation measures include regulatory, legal, economic, financial, information, and voluntary incentives to encourage innovative technological and architectural decisions, the development of new technologies, products and behaviour, innovation in energy saving, renewable energy resources, responsible consumption, and waste management.

Climate change mitigation is one of the key challenges for sustainable development, and the implementation choice of particular climate change policies and measures must be based on the impact of these policies on sustainable development and the climate change policies must be harmonised with sustainable development policies; otherwise, climate change mitigation can bring even more negative effects than the climate change itself. This is particularly the case with developing countries, which are most vulnerable to climate change and have limited possibilities to adapt to climate change. Therefore, climate change mitigation policies in these countries need to be incorporated into the national sustainable development strategies, and their implementation should not hinder the sustainable development goals.

Even if policies and efforts to reduce emissions are successful, some climate change is inevitable. Therefore, the actions to adapt to climate change in Europe and particularly outside the European Union are needed, as the least developed countries are most vulnerable as they have the least financial and technical capacity to adapt.

In order to adapt to climate change, rapid recovery and costs reduction actions need to be taken. Although it is no longer possible to avoid climate change in the next two decades, it is still possible to protect society and economy from these changes to a certain level, for example by providing proper information, improving planning, developing relevant industries, and growing crop better adjusted to the particular climate, as well as improving the entire infrastructure. Adaptation planning in agriculture needs detailed information on regional impacts and meaningful assessment of the adaptive options and their feasibility at the local level. It is essential to improve and refine the spatial and time scales of the assessments of expected climatic impacts and vulnerability, to better understand the climate change impact, and to find the best opportunities to adapt to this change.

References

À propos de l'ISO. 2014. http://www.iso.org/iso/fr/home/about.htm (accessed May 5, 2020).

Adger, W. N. 2003. Social capital, collective action, and adaptation to climate change. *Economic Geography*, 79(4): 387–404.

Ahmad, Q. K., Warrick, R. A., Downing, T. E., Nishioka, S., Parikh, K.S., Parmesan, C., Schneider, S. H., Toth, F. and G. Yohe. 2001. Methods and tools. In: *Climate Change 2001: Impacts, Adaptation, and Vulnerability. Contribution of Working Group II to the Third Assessment Report of the Intergovernmental Panel on Climate Change*, McCarthy, J. J., Canziani, O. F., Leary, N. A., Dokken, D. J. and K. S. White, Eds., Cambridge University Press, Cambridge, 105–143.

Ajibade, I., McBean, G. and Bezner-Kerr, R. 2013. Urban flooding in Lagos, Nigeria: Patterns of vulnerability and resilience among women. *Global Environmental Change*, 23(6): 1714–1725.

Amengual, A., Homar, V., Romero, R., Brooks, H., Ramis, C., and Alonso, S. 2014. Projections of heat waves with high impact on human healthin Europe. *Global and Planetary Change*, 3: 71–84.

Archer, D. 2007. *Global Warming: Under-Standing the Forecast*. Blackwell, Malden, MA.

Arrow, K. J. et al. 2012. Sustainability and the measurement of wealth. *Environment and Development Economics*, 17(3): 317–353.

Azevedo de Almeida, B. and Mostafavi, A. 2016. Resilience of infrastructure systems to sea-level rise in coastal areas: Impacts, adaptation measures, and implementation challenges. *Sustainability*, 8(11): 1115.

Baldwin, R. and Cave, M. 1999. *Understanding Regulation. Theory, Strategy and Practice.* Oxford University Press, Oxford.

Baležentis, T., Tianxiang, L. and Štreimikienė, D. 2016. Is the Lithuanian economy approaching the goals of sustainable energy and climate change mitigation? *Journal of Cleaner Production*, 116(2016): 23–31.

Bassi, S., Dechezleprete, A. and Fankhauser, S. 2013. Climate change policies and the UK business sector: Overview, impacts and suggestions for reform. http://www.lse.ac.uk/GranthamInstitute/wp-content/uploads/2013/11/Climate-change-policies-and-the-UK-business-sector-Nov2013.pdf (accessed May 5, 2020).

Baumilienė, V. 2001. Economic globalization: Development factors and contradictions. Economics and management. Kn.3. *Proceedings of the International Conference.* Kaliningrad State Technical University. Kaunas, Technologija, 362.

Bell, M. L., McDermott, A., Zeger, S. L., Samet, J. M., and Dominici, F. 2004. Ozone and short-term mortality in 95 US urban communities, 1987–2000. *JAMA*, 292(19): 2372–2378.

Bell, S., Alves, S., Oliveira, E. S. and Zuin, A. 2010. Migration and land use change in Europe: A review. *Living Reviews in Landscape Research*, 2. doi: 10.12942/lrlr-2010-2.

Bicknell, J., Dodman, D. et al., Eds. (2009). *Adapting Cities to Climate Change: Understanding and Addressing the Development Challenges.* Earthscan, London. doi: 10.1080/17535069.2013.846001 (accessed May 5, 2020).

Bineswaree, B. and Freund, C. 2004. Trade regulations, and growth. *Trade and Growth Conference*, Research Department Hosted by the International Monetary Fund Washington, DC, January 9, 2006.

BMU. 2016. Klimaschutzplan 2050 Klimaschutzpolitische Grundsätze und Ziele der Bundesregierung. Klimaschutzplan 2050, https://www.bmu.de/fileadmin/Daten_BMU/Download_PDF/Klimaschutz/klimaschutzplan_2050_bf.pdf (accessed May 5, 2020).

Brazauskienė, B. and Kazlauskienė, A. 2002. Labor problems in the conditions of globalization. *Economics and Management: Proceedings of an International Scientific Conference*. Kaunas Technology.

Brooks, I., Weatherston, J., Wilkinson, G. 2010. *The International Business Environment: Challenges and Changes*. Financial Times/Prentice Hall, New York. 347 p.

Carbon Brief. 2018. Q&A: How will China's new carbon trading scheme work? https://www.carbonbrief.org/qa-how-will-chinas-new-carbon-trading-scheme-work žiūrėta (accessed May 5, 2020).

Carbon Dioxide Information Analysis Center (CDIAC), prieiga per internet. http://cdiac.ornl.gov/CO2_Emission/ (accessed May 5, 2020).

Cascio, J. 1999. *The ISO 14000 Handbook*. ASQ Quality Press, Milwaukee, WI.

Chakori, S. 2017. Building a sustainable society: The Necessity to change the Term consumer. *Interdisciplinary Journal of Partnership Studies*, 4(3), Article 9, http://pubs.lib.umn.edu/ijps/vol4/iss3/9 (accessed May 5, 2020).

Čiarnienė, R. and Kumpikaitė, V. 2011. International labour migration. *Students Viewpoint*, 22(5): 527–533.

Čiarnienė, R., Kumpikaitė, V. and Taraškevičius, A. 2009. Influence of macroeconomical indicators on migration process: Theoretical and practical aspects. *Economics and Management*, 14: 553–559.

Čiegis, R. 2004. *Economy and Environment: Managing Sustainable Development. Monograph*. Vytautas Magnus University (Vytautas Magnus University Publishing House), Kaunas, 552.

Čiegis, R. 2006. Ecological Security: New Challenges for the Planet. Strategic Self-Government. Vilnius University Publishing House, Klaipeda, 1(3): 22–33.

Climate Action Tracker, CAT. 2018. Climate action tracker. https://climateactiontracker.org/. (accessed May 5, 2020).

Copeland, B. R. and Taylor, M. S. 1994. North-South trade and the environment. *The Quarterly Journal of Economics*, August: 755–787.

Dagher, G. K. and Itani, O. 2014. Factors influencing green purchasing behavior: Empirical evidence from the Lebanese consumers. *Journal of Consumer Behaviour*, 13(195): 188–195.

Daly, H. 1980. *Towards a Steady State Economy*. W.H. Freeman and Co, San Francisco, 372.

Daly, H. 1991. *Steady State Economics*. Island Press, Washington, DC. 2nd edition, 302.

Damulienė, A. 2013. The problem of migration in Lithuania and its impact on the country's economy. *Business Systems and Economics*, 3(1): 106–118.

Dankelman, I., Alam, K., Ahmed, W. B., Gueye, Y. D., Fatema, N., and Mensah-Kutin, R. 2008. Gender, climate change and human security lessons from Bangladesh, Ghana and Senegal. In The Women's Environment and Development Organization (WEDO) with ABANTU for Development in Ghana, ActionAid Bangladesh and ENDA in Senegal.

Dasgupta, P. 2000. Economic progress and the idea of social capital. In: *Social Capital: A Multifaceted Perspective*, P. Dasgupta, and I. Serageldin, eds., World Bank, Washington, DC.

Dasgupta, P. 2003. Social capital and economic performance: Analytics. In: *Critical Studies in Economic Institutions: Foundations of Social Capital*, E. Ostrom, and T.K. Ahn, eds., Foundations of Social Capital, Edward Elgar, Cheltenham, pp. 309–339

Delventhal, S. 2016. The 10 fastest growing green startups in 2016. https://www.investopedia.com/articles/investing/021116/10-fastest-growing-green-startups-2016.asp (accessed May 5, 2020).

Department of Energy. 2018. Industrial energy efficiency. https://www.savingenergy.org.za/programmes/ (accessed May 5, 2020).

Dimitrov, R. 2016. The Paris agreement on climate change: Behind closed door. https://politicalscience.uwo.ca/people/faculty/full-time_faculty/GEP%20Paris%20Agreement.pdf (accessed May 5, 2020).

Donkor, E. A., Mazzuchi, T. A., Soyner, R., and Roberson, J. A. 2014. Urban water demand forecasting—Review of methods and models. *Journal of Water Resources Planning and Management*, 140: 146–159.

Douglas, S. M., Arango, J., Hugo, G., Kouaouci, A., Pellegrino, A. and Taylor, J. E. 2011. Theories of International migration: A review and appraisal. *Population and Development Review*, 19(3): 431–466.

Dreher, A., Gaston, N., Martens, P. 2008. *Measuring Globalization: Gauging Its Consequences*. Springer Verlag, New York. 224 p.

Economic Migration Regulation Strategy of the Republic of Lithuania. 2007. https://splash-db.eu/policydocument/economic-migration-regulation-strategy/ (accessed May 10, 2020).

EDGAR (Emissions Database for Global Atmospheric Research) (JRC/PBL). 2016. http://www.fao.org/faostat/en/?#data/ (accessed May 5, 2020).

Ehrlich, P. R., Ehrlich, A. H. 1989. *How the Rich Can Save the Poor and Themselves: Lessons from the Global Warming*. Stanford University, Stanford, CA. Paper number 0015.

Eisenhammer, S. 2017. Brazil launches database to fight illegal Amazon logging. https://www.reuters.com/article/us-brazil-environment-amazon-idUSKBN16E2O9 (accessed May 5, 2020).

EM-DAT. 2020. https://www.emdat.be/ (accessed May 5, 2020).

Enerdata. 2018. Global energy and CO_2 data. https://www.enerdata.net (accessed May 5, 2020).

European Commission and Eurostat. 1999. Towards environmental pressure indicators for the EU. Brussels. First edition. https://op.europa.eu/en/publication-detail/-/publication/80ce03d9-fc51-4bb6-a6fc-c3997efa3525 (accessed May 5, 2020).

European Commission and Eurostat. 2001. Towards environmental pressure indicators for the EU. Brussels. Second edition. https://ec.europa.eu/eurostat/web/products-statistical-books/-/KS-36-01-677 (accessed May 5, 2020).

European Commission. 2019. Facts and figures. http:/ec.europa.eu/environment/ecolabel/facts-and-figures.html (accessed May 5, 2020).

Eurostat. 2001. *Economy-Wide Material Flow Accounts and Derived Indicators. A Methodological Guide*. Statistical Office of the European Union, Luxembourg.

Fankhauser, S. 2013. A Practitioner's Guide to a Low-Carbon Economy: Lessons from the UK. *Climate Policy* 13(3): 345–362.

Fankhauser, S. and Burton, I. 2011. Spending adaptation money wisely. *Climate Policy*, 11(3):1037–1049.
Finnveden, G. and Moberg, Å. 2005. Environmental systems analysis: An overview. *Journal of Cleaner Production*, 13(12): 1165–1173.
Fischer-Kowalski, M. and Hüttler, W. 1998. Society's metabolism. The intellectual history of material flow analysis, part II, 1970–1998. *Journal of Industrial Ecology*, 2(4): 107–136.
Fleming, J. R. 1998. *Historical Perspectives on Climate Change*. Oxford University Press, Oxford.
Fontinelle, E. 2010. 7 ways climate change affects companies. https://www.investopedia.com/financial-edge/0210/7-ways-climate-change-affects-companies.aspx (accessed May 5, 2020).
Franceschini, F., Galetto, M., and Cecconi, P. 2006. A worldwide analysis of ISO 9000 standard diffusion, Benchmarking. *An International Journal*, 13(4): 523–541.
Franchini, M. and Mannucci, P. M. 2015. Impact on human health of climate changes. *European Journal of Internal Medicine*, 26(1): 1–5.
Gallastegui, I. G. 2002. The use of eco-labels: A review of literature. *European Environment*, 12(331): 316–331.
Gartland, L. 2011. *Heat Islands: Understanding and Mitigating Heat in Urban Areas*. Earthscan, London, 192 p.
Gerlagh, R., Dellink, R., Hofkes, M., Verbruggen, H. 2002. A measure of sustainable national income for the Netherlands. *Ecological Economics*, 41: 157–174.
Glacken, C. J. 1967. *Traces on the Rhodian Shore. Nature and Culture in Western Thought from Ancient Times to the End of the Eighteenth Century*. Cambridge University Press, Berkeley, CA.
Global CO_2 concentrations (parts per million). 1980–2016. NOAA. Prieiga per internet. www.esrl.noaa.gov/gmd/ccgg/trends/ (accessed May 5, 2020).
Global Report on Internal Displacement. 2019. http://www.internal-displacement.org/global-report/grid2019/ (accessed May 5, 2020).
Gough C., Castells N., Funtowicz S. 1998. Integrated assessment: An emerging methodology for complex issues. *Environmental Modeling and Assessment*, 3: 19–29.
Government of Canada. 2020. https://www.canada.ca/en.html (accessed May 5, 2020).
Government of Indonesia. 2020. http://www.indonesia.cz/the-government-of-the-republic-of-indonesia/ (accessed May 5, 2020).
Government of Japan. 2020. https://www.japan.go.jp/ (accessed May 5, 2020).
Government of South Africa. 2020. https://www.gov.za/welcome-official-south-african-government-online-site (accessed May 5, 2020).
Greening, L. A. and Bernow, S. 2004. Design of coordinated energy and environmental policies: Use of multi-criteria decision-making. *Energy Policy*, 32(6): 721–735.
Gütschow, J., Jeffery, L., Gieseke, R., Gebel, R. 2018. The PRIMAP-hist national historical emissions time series (1850–2015). http://dataservices.gfz-potsdam.de/pik/showshort.php?id=escidoc:2959897 (accessed May 5, 2020).
Guy, J. A. and Sander, N. 2014. Quantifying global international migration flows. *Science*, 343(6178). https://www.researchgate.net/publication/261186924_Quantifying_Global_International_Migration_Flows (accessed January 15, 2020).
Gylys, P. 2008. *Economics, Anti-Economics and Globalization*. Vilnius University Publishing House, Vilnius. 443 p.

Hallegatte, S. et al. 2013. Future flood losses in major coastal cities. *Nature Climate Change*, 3: 802–806.

Hardin, G. 1968. The tragedy of the commons. *Science*, 162(3859): 1243–1248.

Hardin, G. 1993. *Living within Limits: Ecology, Economics, and Population Taboos*. Oxford University Press, New York, 339.

Hawkes, J. 2001. The fourth pillar of sustainability culture's essential role in public planning. Cultural Development Network (Vic) https://www.researchgate.net/publication/200029531_The_Fourth_Pillar_of_Sustainability_Culture's_essential_role_in_public_planning (accessed January 5, 2021).

Heakel, M. 2015. What Is International Trade? https://www.investopedia.com/articles/03/112503.asp (accessed May 5, 2020).

Herendeen, R. 2004. Energy analysis and emergy analysis: A comparison. *Ecological Modelling*, 178: 227–237.

Howlett, M. 2009. Governance modes, policy regimes and operational plans: A multi-level nested model of policy instrument choice and policy design. *Policy Sciences*, 42(1): 73–89.

Huq, S., Kovats, S., Reid, H. and Satterthwaite, D. 2007. Reducing risks to cities from disasters and climate change. *International Institute for Environment and Development (IIED)*, 19(1): 3–15. doi: 10.1177/0956247807078058.

IEA. 2016. Energy access database. https://www.iea.org/countries (accessed May 5, 2020).

IEA. 2018. Global EV outlook 2018. https://www.iea.org/gevo2018/ (accessed May 5, 2020).

Ihemezie, E. J., Ukwuaba, I. C. and Nnaji, A. P. 2018. Impact of 'green' product label standards on consumer behavior: A systematic review analysis. *International Journal of Academic Research in Business and Social Sciences*, 8(9): 666–684.

International Organization for Migration (IOM). 2014. https://www.iom.int/ (accessed May 5, 2020).

IPCC. 2007. Climate change 2007: The physical science basis. *Contribution of Working Group I to the Fourth Assessment Report of the Intergovernmental Panel on Climate Change*. Solomon, S., D. Qin, M. Manning, Z. Chen, M. Marquis, K.B. Averyt, M. Tignor and H.L. Miller (eds.). Cambridge University Press, Cambridge, United Kingdom and New York.

IPCC. 2014a. Niang, I., O.C. Ruppel, M.A. Abdrabo, A. Essel, C. Lennard, J. Padgham, and P. Urquhart (2014). Africa. In: Climate Change 2014: Impacts, Adaptation and Vulnerability. Part B: Regional Aspects, Chapter 22 (Africa). Contribution of Working Group II to the Fifth Assessment Report of the Intergovernmental Panel on Climate Change. pp. 1199–1265.

IPCC. 2014b. Reisinger, A., R. L. Kitching, F. Chiew, L. Hughes, P. C. D. Newton, S. S. Schuster, A. Tait, and P. Whetton (2014). Australia. In: Climate Change 2014: Impacts, Adaptation and Vulnerability. Part B: Regional Aspects, Chapter 25 (Australasia). Contribution of Working Group II to the Fifth Assessment Report of the Intergovernmental Panel on Climate Change. pp. 35–94.

IPCC. 2014c. Kovats, R. S., R. Valentini, L. M. Bouwer, E. Georgopoulou, D. Jacob, E. Martin, M. Rounsevell, and J.-F. Soussana (2014). Europe. In: Climate Change 2014: Impacts, Adaptation and Vulnerability. Part B: Regional Aspects, Chapter 23 (Europe). Contribution of Working Group II to the Fifth Assessment Report of the Intergovernmental Panel on Climate Change. pp. 1267–1326.

IPCC. 2014d. Magrin, G. O., J.A. Marengo, J.-P. Boulanger, M. S. Buckeridge, E. Castellanos, G. Poveda, F. R. Scarano, and S. Vicuña (2014). Central and South America. In: Climate Change 2014: Impacts, Adaptation and Vulnerability. Part B: Regional Aspects, Chapter 27 (Central and South America). Contribution of Working Group II to the Fifth Assessment Report of the Intergovernmental Panel on Climate Change. pp. 1499–1566.

IPCC. 2014e. Romero-Lankao, P., J. B. Smith, D. J. Davidson, N. S. Diffenbaugh, P. L. Kinney, P. Kirshen, P. Kovacs, and L. Villers-Ruiz (2014). North America. In: Climate Change 2014: Impacts, Adaptation and Vulnerability. Part B: Regional Aspects, Chapter 26 (North America). Contribution of Working Group II to the Fifth Assessment Report of the Intergovernmental Panel on Climate Change. pp. 1439–1498.

IPCC. 2014f. Larsen, J.N., O.A. Anisimov, A. Constable, A. B. Hollowed, N. Maynard, P. Prestrud, T. D. Prowse, and J. M. R. Stone (2014). Polar regions. In: Climate Change 2014: Impacts, Adaptation and Vulnerability. Part B: Regional Aspects, Chapter 28 (Polar Regions). Contribution of Working Group II to the Fifth Assessment Report of the Intergovernmental Panel on Climate Change. pp. 1567–1612.

IPCC. 2014g. Nurse, L. A., R. F. McLean, J. Agard, L. P. Briguglio, V. Duvat-Magnan, N. Pelesikoti, E. Tompkins, and A. Webb (2014). Small islands. In: Climate Change 2014: Impacts, Adaptation and Vulnerability. Part B: Regional Aspects, Chapter 29 (Small Islands). Contribution of Working Group II to the Fifth Assessment Report of the Intergovernmental Panel on Climate Change. pp. 1613–1654.

IPCC. 2018. Global Warming of 1.5°C. An IPCC Special Report on the impacts of global warming of 1.5°C above pre-industrial levels and related global greenhouse gas emission pathways, in the context of strengthening the global response to the threat of climate change, sustainable development, and efforts to eradicate poverty. Masson-Delmotte, V., P. Zhai, H.-O. Pörtner, D. Roberts, J. Skea, P.R. Shukla, A. Pirani, W. Moufouma-Okia, C. Péan, R. Pidcock, S. Connors, J.B.R. Matthews, Y. Chen, X. Zhou, M.I. Gomis, E. Lonnoy, T. Maycock, M. Tignor, and T. Waterfield (eds.). In Press.

Ireland, G. and Burton, J. 2018. *An Assessment of New Coal Plants in South Africa's Electricity Future: The Cost, Emissions, and Supply Security Implications of the Coal IPP Programme.* Energy Research Centre, University of Cape Town, Cape Town.

Kann, A. and Weyant J. P. 2000. Approaches for performing uncertainty analysis in large-scale energy/economic policy models. *Environmental Modeling and Assessment*, 5: 29–46.

Kauggman, C. and Tebar, C. 2009. Business and climate change: An MNE guidelines perspective. https://www.oecd.org/daf/inv/corporateresponsibility/43356899.pdf (accessed May 5, 2020).

Kelly, P.M. and Adger, W.N. 2000. Theory and practice in assessing vulnerability to climate change and facilitating adaptation. *Climatic Change* 47: 325–352.

Keohane, R. O. 1999. International relations and international law: Interests, reputation, institutions. *Proceedings of the Annual Meeting (American Society of International Law)*, 93: 375–379 Published by Cambridge University Press on behalf of the American Society of International Law https://www.jstor.org/stable/25659335 (accessed May 5, 2020).

Keohane, R. O. 2003. The Concept of Accountability in World Politics and the Use of Force. *Michigan Journal of International Law* 24(4): 1–21.

Kim Chang-gil. 2009. Strategies for implementing green growth in agricultural sector. In *Proceedings in Green Korea 2009 - Green Growth and Cooperation. National Research Council for Economics, Humanities and Social Science.*

Kleijn, R. 2001. Adding it all up. The sense and non-sense of bulk-MFA. *Journal of Industrial Ecology*, 4(2): 7–8.

Knowlton, K. et al. 2014. Development and Implementation of South Asia's First Heat-Health Action Plan in Ahmedabad. *International Journal of Environmental Research and Public Health*, 11(4): 3473–3492. doi: 10.3390/ijerph110403473.

Krugman, P. 1991. Increasing returns and economic geography. *Journal of Political Economy*, 99(3): 483–499.

Laczko, F. and Aghazarm, C. 2009. Migration, environment and climate change: Assessing the evidence.Geneva, Switzerland. *International Organization for Migration*, 448. https://publications.iom.int/books/migration-environment-and-climate-change-assessing-evidence (accessed May 5, 2020).

Lammers, P. E. M. and Gilbert, A. J. 1999. *Towards Environmental Pressure Indicators for the EU: Indicator Definition*. EUROSTAT, Brussels.

Levitt, T. 1983. The globalization of markets. *Harvard Business Review*, 61: 92–102. ISBN 978-0-000-83308-2.

Lindqvist, A. and von Malmborg F. 2004. What can we learn from local flow analyses? The review of cadmium flows in Swedish municipalities. *Journal of Cleaner Production*, 12: 909–918.

Maier, S. 2005. *Do Good Environmental Management Systems Lead to Good Environmental Performance?* EIRIS Foundation, London.

Marcus, P. A. and Willing, J. T. 1997. Moving ahead with ISO 14000: Improving management and advancing sustainable development.

Matthews, E., Bringezu, S., Fischer-Kowalski, M., Huetller, W., Kleijn, R., Moriguchi, Y., Ottke, C., Rodenburg, E., Rogich, D., Schandl, H. et al. 2000. *The Weight of Nations. Material Outflows from Industrial Economies*, World Resource Institute, Washington, DC.

McGranahan, G., Balk, D. and Anderson, B. 2007. The rising tide: Assessing the risks of climate change and human settlements in low elevation coastal zones. *Environment and Urbanization*, 19(1): 17–37.

Meadows, D. H., Meadows, D. L., Randers J. et al. 1972. *Limits to Growth*. Universe Books, New York, 205.

Mendelsohn, R. 2012. Development and climate adaptation. In R. Hahn & A. Ulph (eds.), *Climate Change and Common Sense: Essays in Honor of Tom Schelling*. Oxford University Press, Oxford.

Mendelsohn, R. and Schlesinger, M. 1999. Climate-response functions. *Ambio*, 28(4): 362–366.

Mendelsohn, R., et al. 2006. The distributional impact of climate change on rich and poor countries. *Environment and Development Economics*, 11(2):159–178.

Metroeconomica. 2004. *Costing the impacts of climate change in the UK: Overview of guidelines*. UKCIP Technical Report. UKCIP, Oxford.

Migration, Environment and Climate Change: Evidence for Policy (MECLEP). 2014. https://environmentalmigration.iom.int/migration-environment-and-climate-change-evidence-policy-meclep (accessed May 5, 2020).

Ministry of Energy and Mineral Resources Republic of Indonesia. 2020. https://www.esdm.go.id/en (accessed May 5, 2020).

MOEF. 2018. Draft national forest policy 2018. http://www.moef.nic.in/sites/default/files/Draft%20National%20Forest%20Policy%2C%202018.pdf

Moen, R., and Norman, C. 2009. The History of the PDCA Cycle. In Proceedings of the 7th ANQ Congress, Tokyo 2009, September 17. https://rauterberg.employee.id.tue.nl/lecturenotes/DG000%20DRP-R/references/Moen-Norman-2009.pdf

Möller, I., Kudella, M., Rupprecht, F., Spencer, T., Paul, M., van Wesenbeeck, B., Wolters, G., Jensen, K., Bouma, T. J., Miranda-Lange, M., and Schimmels, S. 2014. Wave attenuation over coastal salt marshes under storm surge conditions. *Nature Geoscience*, 7 (Oct 2014). doi: 10.1038/NGEO2251.

Mongabay. 2018. Indonesian president signs 3-year freeze on new oil palm licenses. https://news.mongabay.com/2018/09/indonesian-president-signs-3-year-freeze-on-new-oil-palm-licenses/ (accessed May 5, 2020).

Munasinghe, M. 1991. Groundwater resource management and environmental protection. A case study of the Philippines. https://doi.org/10.1111/j.1477-8947.1991.tb00150.x (accessed May 5, 2020).

National Climate Change Response White Paper. 2011. https://www.environment.gov.za/sites/default/files/legislations/national_climatechange_response_whitepaper.pdf (accessed May 10, 2020).

National Research Council. 2016. Assessment of approaches to updating the social cost of carbon: Phase 1 report on a near-term update. National Academies Press, Washington, DC.

Newell, P. 2005. Climate for change? Civil society and the politics of global warming. *Global Civil Society*, 6: 1–31.

Nordhaus W. 2014. Estimates of the social cost of carbon: Concepts and results from the DICE-2013R model and alternative approaches. *Journal of Environmental and Resource Economics*, 1(1/2): 273–312.

OECD. 2018. *Cost-Benefit Analysis and the Environment: Further Developments and Policy Use*. OECD Publishing, Paris, https://doi.org/10.1787/9789264085169-en. (accessed May 5, 2020).

Our Common Future. 1987. World Commission on Environment and Development. https://sustainabledevelopment.un.org/content/documents/5987our-common-future.pdf (accessed May 5, 2020).

Our World in Data. 2020. https://ourworldindata.org/natural-disasters (accessed May 5, 2020).

Papademetriou, D. G. and Martin, P.L. 1991. *The Unsettled Relationship: Labor Migration and Economic Development*. Green-wood Press, New York, 3–26.

PricewaterhouseCoopers. 2004. Global annual review. https://s3-us-west-2.amazonaws.com/ungc-production/attachments/7107/original/PwC_Global_Annual_Review_2004.pdf?1282019080 (accessed May 5, 2020).

Reardon, T. 2016. Growing Food for Growing Cities: Transforming Food Systems in an Urbanising World. The Chicago Council on Global Affairs, Chicago.

Report by the Shelton Group. 2013. The social norm tipping point. *Green Living Pulse*, 12 http://sheltonsites.com/ sheltongrp1/files/2012/10/GreenLivingPulse2012_executive-summary.pdf. (accessed May 1, 2020)..

Report of the United Nations Conference on the Human Environment, Stockholm. 5–16 June 1972 https://digitallibrary.un.org/record/523249#record-files-collapse-header (accessed May 7, 2020).

Revi, A., Satterthwaite, D., Aragón-Durand, F., Corfee-Morlot, J., Kiunsi, R. B. R., Pelling, M., Roberts, D. C., and Solecki, W. 2014. Urban areas. In: Field, C. B., Barros, V. R., Dokken, D. J., Mach, K. J., Mastrandrea, M. D., Bilir, T. E., Chatterjee, M., Ebi, K. L., Estrada, V., Genova, R. C., Girma, B., Kissel, E. S., Levy, A. N., Maccracken, S., Mastrandrea, P. R., and White, L. L. (eds.), *Climate Change 2014: Impacts, Adaptation, and Vulnerability. Part A: Global and Sectoral Aspects. Contribution of Working Group II to the Fifth Assessment Report of the Intergovernmental Panel on Climate Change.* Cambridge University Press, Cambridge, 535–612.

Richard, G., Newell, W. A. and Aimi, D. 2013. Carbon markets 15 years after Kyoto: Lessons learned, new challenges. *Journal of Economic Perspectives*, 27(1): 123–146. http://pubs.aeaweb.org/doi/pdf/10.1257/jep.27.1.123 (accessed May 5, 2020).

Rigaud, K. K., de Sherbinin, A., Bryan, J., Bergmann, J., Clement, V., Ober, K., Schewe, J., Adamo, S., McCusker, B., Heuser, S., Midgley, A. 2018. *Groundswell: Preparing for Internal Climate Migration*. World Bank, Washington, DC. © World Bank. https://openknowledge.worldbank.org/handle/10986/29461 (accessed May 5, 2020).

Rogel, L., Elzen, M., Hohne, N. and Fransen, T. 2016. Paris Agreement climate proposals need a boost to keep warming well below 2°C. https://www.ncbi.nlm.nih.gov/pubmed/27357792 doi: 10.1038/nature18307 (accessed May 5, 2020).

Rosen, M. and Dincer I. 2001. Exergy as the confluence of energy, environment and sustainable development. *Exergy International Journal*, 1(1): 3–13.

Rosenzweig, C., Solecki, W., Romero-Lankao, P., Mehrotra, S., Dhakal, S., & Ali Ibrahim, S. (eds.). 2018. *Climate Change and Cities: Second Assessment Report of the Urban Climate Change Research Network*. Cambridge University Press. New York.

Rudžinskienė, R. and Paulauskaitė, L. 2014. Causes of Lithuanian emigration and its consequences for the country's economy. *Social Theory, Empiricism, Policy and Practice*, 8: 62–78.

Ruth, M., Bernier, C., Jollands, N., and Golubiewski, N. 2007. Adaptation of urban water supply infrastructure to impacts from climate and socioeconomic changes: The case of Hamilton. *New Zealand: Water Resources Management*, 21: 1031–1045.

Ruževičius, J. 2007. *Quality Management Methods and Modules*. Vilnius University, Vilnius.

Salinas, G. and Aksoy, A. 2006. Growth before and after trade liberalization. World Bank Policy Research Working Paper 4062, November 2006.

Santos, M. A. O. 2007. Minimizing the business impact on the natural environment: A case study of Woolworths South Africa. *European Business Review*, 23(4): 384–391.

Scanlon, B. R., Duncan, I., and Reedy, R. C. 2013b. Drought and the water-energy nexus in Texas. *Environmental Research Letters*, 8(4): 045033.

Scanlon, B. R., Reedy, R. C., Duncan, I., Mullican, W. F., and Young, M. 2013a. Controls on water use for thermoelectric generation: Case study Texas, U.S. *Environmental Science & Technology*, 47(19): 11326–11334.

Schelling, Thomas C. 1992. Some Economics of Global Warming. American Economic Review. *American Economic Association* 82(1): 1–14.

Schleich, J. and Hillenbrand, T. 2009. Determinants of residential water demand. *Ecological Economics*, 68: 1756–1769.

Scott, L. and Vigar-ellis, D. 2014. Consumer understanding, perceptions and behaviours with regard to environmentally friendly packaging in a developing nation. *International Journal of Consumer Studies*, 38(6): 642–649.

Segbers, K. and Imbrusch, K. 2000. The globalization of Eastern Europe: Teaching international relations without borders. In: *Hamburg*, 587 p. ISSN 3825847292.

Shah, R. 2018. Government finally wakes up: Sets a realistic goal of 30% electric vehicles by 2030 from existing 100% target. https://www.financialexpress.com/auto/car-news/government-finally-wakes-up-sets-a-realistic-goal-of-30-electric-vehicles-by-2030-from-existing-100-target/1091075/ (accessed May 5, 2020).

Smith, L. 2016. What does it mean to be green? https://www.investopedia.com/articles/basics/07/green-investing.asp (accessed May 5, 2020).

Staniškis, J. and Stasiškienė, Ž., 2007. Industrial production development and implementation system for cleaner production (APINI-SPIN). *Environmental Research, Engineering and Management*, 4(42): 54–59.

Stanway, D. 2018. China to create new forests covering area size of Ireland. https://www.reuters.com/article/us-china-environment-forest/china-to-create-new-forests-covering-area-size-of-ireland-china-daily-idUSKBN1EU02L (accessed May 5, 2020).

Stern, N. H. 2007. *The Economics of Climate Change: the Stern Review*. Cambridge University Press, Cambridge.

Stockwell, E. G., Groat, H. T. 1984. *Word Population. An Introduction to Demography*. Franklin Watts, New York, London, Toronto.

Stoker, P. and Rothfeder, R. 2014. Drivers of urban water use. *Sustainable Cities and Society*, 12: 1–8.

Streimikiene, D. 2005. Indicators for sustainable energy development in Lithuania. *Natural Resources Forum*, 29(4). doi:10.1111/j.1477-8947.2005.00144.x.

Štreimikienė, D. 2010. External costs of energy security and climate change. *International Energy Journal*, 11: 225–234.

Sullivan, L. 2013. Climate change strategy for business. https://www.theguardian.com/sustainable-business/blog/climate-change-strategy-business-strategic-priority (accessed May 5, 2020).

Tang, H. and Harrison, A. 2004. Liberalization of trade: Why so much controversy? World Bank. www1.worldbank.org/prem/lessons1990s/Chap%205%20trade%20092104%20rw.pdf Online at http://mpra.ub.uni-muenchen.de/36280/ MPRA Paper No. 36280, posted 29. February 2012.

Testa, F., Iraldo, F., Vaccari, A., and Ferrai, E. 2015. Why eco-labels can be effective marketing tools: Evidence from a study on Italian consumers. *Business Strategy and Environment*, 24(4): 252–265.

The World Bank. 2020. https://data.worldbank.org/indicator/SM.POP.TOTL.ZS (accessed May 5, 2020).

The World Bank. 2020. https://data.worldbank.org/topic/trade (accessed May 5, 2020).

The World Bank. 2020. https://data.worldbank.org/indicator/SP.POP.1564.TO.ZS (accessed May 5, 2020).

Timberley, J. 2018. The carbon brief profile: Brazil. https://www.carbonbrief.org/the-carbon-brief-profile-brazil (accessed May 5, 2020).

Tol, R. S. J., et al. 2004. Distributional aspects of climate change impacts. *Global Environmental Change*, 14(3): 259–272.

Tompkins, E. L., Mensah, A., King, L., Long, T. K., Lawson, E. T., Hutton, C., Hoang, V. A., Gordon, C., Fish, M., Dyer, J., and Bood, N. 2013. An investigation of the evidence of benefits from climate compatible development. Sustainability Research Institute.

TransportPolicy.net. 2018. Indonesia: Lightduty emissions. https://www.transportpolicy.net/standard/indonesia-light-duty-emissions/?title=indonesia:_lightduty (accessed May 5, 2020).

Tyler, S., and Moench, M. 2012. A framework for urban climate resilience. *Climate and Development*, 4: 311–326.

UCCRN Technical Report, 2018. How climate change could impact the world's greatest cities. https://cittaclima.it/wp-content/uploads/2018/06/1789_Future_We_Dont_Want_Report_1.4_hi-res_120618.original.pdf (accessed January 5, 2021).

UN World Urbanization Prospects. 2018. Department of Economic and Social Affairs United Nations New York, 2019 https://population.un.org/wup/Publications/Files/WUP2018-Report.pdf (accessed May 5, 2020).

UNCSD. 2001. *Indicators of Sustainable Development: Guidelines and Methodologies*. United Nations, New York.

UNESCO. 2005. Cultural diversity and globalization. http://unesdoc.unesco.org/images/0013/001393/139318e.pdf (accessed May 5, 2020).

UN-Habitat. 2013. Planning and design for sustainable urban mobility. Global Report on Human Settlements. UN-Habitat. https://unhabitat.org/urban-themes/energy/ (accessed May 5, 2020).

United Nations Population Fund. 2007. https://www.unfpa.org/ (accessed May 10, 2020).

United Nations. 2002. National implementation of agenda 21: A report. New York, Department of Economic and Social Affairs, Division for Sustainable Development, National Information Analysis Unit, United Nations.

United Nations. 2017. Sustainable development goals report. https://www.un.org/development/desa/publications/sdg-report-2017.html (accessed May 5, 2020).

Urbonaitė, E. 2007. Economic migration policy. Geopolitics. http://www.geopolitika.lt/?artc=964 (accessed May 5, 2020).

Ürge-Vorsatz, D., Herrero, S. T., Dubash, N. K. and Lecocq, F. 2014. Measuring the co-benefits of climate change mitigation. *Annual Review of Environment and Resources*, 39: 549–582.

Vicuña, S., Redwood, M., Dettinger, M., and Noyola, A. 2018. Urban water systems. In Rosenzweig, C., W. Solecki, P. Romero-Lankao, S. Mehrotra, S. Dhakal, and S. Ali Ibrahim (eds.), *Climate Change and Cities: Second Assessment Report of the Urban Climate Change Research Network*. Cambridge University Press, New York, 519–552.

von Weizsäcker, E. and Jesinghaus, J. 1992. *Ecological Tax Reform. Policy Proposal for Sustainable Development*. Zed Books, London.

Wilhelmi, O. V., Purvis, K. L., & Harriss, R. C. 2004. Designing a geospatial information infrastructure for mitigation of heat wave hazards in urban areas. *Natural Hazards Review*, 5(3): 147–158.

Winters, A. L., McCulloch, N. and McKay, A. 2004. Trade liberalization and poverty: The evidence so far. *Journal of Economic Literature*, XLII: 72–115.

World Bank. 2006. *Investment Framework for Clean Energy and Development*. World Bank, Washington, DC.

World Bank. 2010. *The Economics of Adaptation to Climate Change*. World Bank, Washington, DC.

World Bank. 2018. Access to electricity. https://data.worldbank.org/indicator/EG.ELC.ACCS.ZS.

World Bank. 2020. World development indicators. https://data.worldbank.org/products/wdi.

World Conservation Strategy. 1980. Living resources conservation for sustainable development. https://portals.iucn.org/library/efiles/documents/wcs-004.pdf (accessed May 5, 2020).

World Summit on Sustainable Development (WSSD). 2002. Published by the International Institute for Sustainable Development (IISD). http://enb.iisd.org/2002/wssd/WSSDcompilation.pdf (accessed May 5, 2020).

Yongqiang, L., Stanturf, J. and Goodrick, S. 2010. Trends in global wildfire potential in a changing climate. *Forest Ecology and Management*, 259(4): 685–697. http://www.sciencedirect.com/science/article/pii/S0378112709006148 (accessed May 5, 2020).

Zhou, S. L., McMahon, T. A., Walton A., and Lewis, J. 2000. Forecasting daily urban water demand—A case study of Melbourne. *Journal of Hydrology*, 236: 153–164.

Index

For Product Safety Concerns and Information please contact our EU representative GPSR@taylorandfrancis.com Taylor & Francis Verlag GmbH, Kaufingerstraße 24, 80331 München, Germany